BASIC CONCEPTS IN
CELL BIOLOGY
AND HISTOLOGY

. . . .

Notice

BASIC CONCEPTS IN CELL BIOLOGY AND HISTOLOGY

A STUDENT'S SURVIVAL GUIDE

James C. McKenzie, Ph.D.
Associate Professor
Department of Anatomy
Howard University College of Medicine
Washington, DC

Robert M. Klein, Ph.D.
Professor, Department of Anatomy and Cell Biology
School of Medicine
University of Kansas Medical Center
Kansas City, Kansas

Series Editor
Hiram F. Gilbert, Ph.D.

McGraw-Hill
Health Professions Division
New York   St. Louis   San Francisco
Auckland   Bogota   Caracas   Lisbon   London   Madrid
Mexico City   Milan   Montreal   New Delhi   Paris   San Juan
Singapore   Sydney   Tokyo   Toronto

McGraw-Hill

A Division of The McGraw-Hill Companies

1234567890 DOCDOC 99

ISBN 0-07-036930-5

This book was set in Times Roman by Better Graphics, Inc. The editors were Joseph Hefta, Susan R. Noujaim, and Peter J. Boyle; series editor was Hiram F. Gilbert, Ph.D.; the production supervisor was Helene G. Landers; the cover designer was Marsha Cohen / Parallelogram. R. R. Donnelley and Sons Company was the printer and binder.

This book is printed on acid-free paper.

Cataloging-in-Publication Data is on file for this book at the Library of Congress.

Basic concepts in cell biology and histology / editors, James C.
   McKenzie, Robert M. Klein.
        p.   cm.
   Includes bibliographical references and
   ISBN 0-07-036930–5. -- ISBN 0-07-036930-5 (alk. paper)
   1. Cytology Outlines, syllabi, etc.   2. Histology Outlines,
syllabi, etc.   I. Klein, Robert M. (Robert Melvin), 1949–
   [DNLM: QS 504 B311 1999]
   QH581.5.B37   1999
   611'.018--dc21
   DNLM/DLC
   for Library of Congress   99-32476
                                                                 CIP

To Drs. John C. Belton and Steve Benson
who made my early forays into histology and cell biology
interesting, exciting, and rewarding;
and to Robert M. Klein, Ph.D.—
educator, mentor, colleague, and
most importantly, friend.
(JCM)

To my parents, Nettie and David Klein,
for building the foundation;
to my wife Beth, for her love and support; and
to my children, Melanie, Jeffrey, and David,
for their interest and
the promise of bright futures.
(RMK)

· C O N T E N T S ·

P A R T   I I
BASIC HISTOLOGY / 145

INDEX   **407**

· P R E F A C E ·

This Basic Concepts text organizes and integrates the immense fields of cell biology and histology. *Basic Concepts in Cell Biology and Histology* is not intended to be a comprehensive textbook, but a systematic compendium of key concepts about the relationship between cells, tissues, and organ systems and the way in which these histological entities function. The first part of the book develops the basic concepts of how cells are organized internally to carry out both generalized and specific functions. The second part of the book focuses on the arrangement of cells to form tissues and organs which work together to carry out organ system functions. Furthermore, the cell biology concepts developed in the first ten chapters are applied to the tissues and organs described in the remaining chapters. The last two chapters of this book serve as a road map to the identification of tissues and organs and a review of the basic techniques used in cell biology and histology. Also included is a guide to the orientation and interpretation of electron micrographs.

Each chapter contains primary topics (bold headings) which are organized into a series of basic concepts (within boxes) proceeding from the most basic to the more specific. To be successful in the study of cell biology and histology it is essential to master these basic concepts. In turn, each basic concepts box is followed by a short paragraph of explanation. Within these paragraphs, important terms are italicized. These are terms that are deemed essential to understanding and discussion of the chapter topic. The arrangement of basic concepts boxes and the highlighting of key terms which comprise the vocabulary of cell biology and histology are intended to facilitate student learning.

The flowcharts provided in the Strategies chapter are a unique portion of the book. They have been developed through the combined experience of the authors: fifty years of teaching over 7000 medical, dental, and graduate students in our respective educational institutions. These flowcharts facilitate the efforts of the beginning student to circumnavigate the obstacle that all tissues and organs appear indiscriminately pink and purple while learning the tried and true methods for their diagnostic identification. The flowcharts are organized to allow histological identification of any organ with an average of only four or five specific decisions, such as "Is this a solid organ or does it have a lumen?" or "What type of epithelium is present?"

· A C K N O W L E D G M E N T ·

The authors gratefully acknowledge their students and colleagues at Howard University College of Medicine and The University of Kansas Medical Center for their insightful comments and cogent criticisms of the manuscript. The following faculty from Howard University are acknowledged for their input and contributions: James H. Baker, William Ball, Robert J. Cowie, Raziel Hakim, and John K. Young. The following faculty from the University of Kansas are acknowledged for their input and contributions: George C. Enders, Lisa K. Felzien, and Vincent H. Gattone, II. In addition, the immunologic expertise of Gary W. Wood at Wayne State University School of Medicine and proofreading of John Clancy, Jr., of Loyola University—Stritch School of Medicine are appreciated. The authors thank David Cormack of the University of Toronto for providing the long bone development figure for Chapter 14. The authors thank the artists who painstakingly converted their ideas and sketches into exquisite drawings. In alphabetical order, the artists involved in completion of the artwork for this book were: John Berman, Andre Branning, Karen Chinn, Larry Howell, and Bill Paige. The authors acknowledge their present and past students who have challenged them to develop new teaching methodologies and whose comments and criticisms have improved the content of this text. Our students have provided our raison d'être for writing *Basic Concepts in Cell Biology and Histology* and have kept us excited about this endeavor. We anticipate that our future students will find this text to be beneficial to their studies of cell biology and histology.

· P A R T · I ·

BASIC CELL BIOLOGY

·

BASIC CELL BIOLOGY

MEMBRANES

·

- **Membrane Lipids**
- **Membrane Proteins**

· · · · · · · · · · · ·

As Allan Sherman once sang

> You gotta have skin.
> All you really need is skin.
> 'Cause skin is the thing
> that keeps your insides in.

Indeed, cell membranes do, to a large extent, keep the insides of a cell in and the outside of the cell out. Membranes also form boundaries of cytoplasmic organelles such as the nucleus, mitochondria, lysosomes, endoplasmic reticulum, Golgi apparatus, and transport and secretory vesicles. However, membranes play a much greater role than merely acting as barriers. First, their composition allows them to be more selective than the bouncer at the local hot nightspot, allowing passage to only the "beautiful" molecules and keeping out the "riff-raff." The beautiful molecules tend to be (1) those which are lipid soluble and can diffuse easily through the membrane and (2) those for which special doors (channels and transporters) exist. Second, membranes also serve as sites for communication with the external milieu via *receptors and recognition molecules* embedded in or connected to their superstructure. Third, membranes are also the location of macromolecules by which cells *anchor* themselves to each other or the extracellular "gunk" (matrix). In addition, they provide attachment sites for contractile or cytoskeletal proteins which alter or maintain cell shape. Finally, membranes are dynamic. Their basic components can change rapidly to meet the demands of life in the fast lane, or their basic morphology can be altered to allow the passage of big packages of "stuff" in or out of the cell. The key to understanding the structure and function of cell membranes is *compartmentalization*.

MEMBRANE LIPIDS

All cellular membranes are based on the lipid bilayer.

Most lipids involved in the formation of cellular membranes are amphipathic, that is, they have split personalities. The portion consisting of long-chain fatty acids composed solely of carbon and hydrogen is very hydrophobic. It wants about as much to do with water as a child at bath time. The other portion of the lipid consists of a charged or polar head group, which likes water. Consequently, when lipids are mixed with water, they arrange themselves so that their polar portions face water and their nonpolar portions face away from water. One way to accomplish this in a water-filled environment is to form two layers of lipid molecules with polar groups on the outer and inner surfaces and nonpolar groups facing each other in the middle (Fig. 1-1).

Cell membranes are dynamic, fluid structures.

The rigid, constrained structure seen when viewing an electron micrograph of a cell membrane is an artifact! It has been treated with chemicals (fixatives like glutaraldehyde and osmium tetroxide which cross-link molecules) and made to stand still to have its picture taken. In the bilayer of a cell membrane, lipid molecules are free to move in several directions. They may rotate about their long axis, jostle laterally with each other, and occasionally flip from one side of the bilayer to the other.

The fluidity of a cell membrane is influenced by its composition.

Two characteristics of the fatty acid components of membrane lipids influence the fluidity of the membrane: *fatty acid chain length* and the *number of double bonds* in the fatty acid chain. Long-chain fatty acids of uniform length pack well together and thus limit movement. The addition of a few shorter fatty acid chains creates nooks and crannies in the bilayer into which other chains can move. The addition of double bonds (*unsaturation*) to the fatty acid chains inhibits axial rotation at the double bonds and induces a kink in the fatty acid chain which inhibits orderly packing and thus promotes movement in the membrane (Fig. 1-1).

Cholesterol also influences the fluidity of cell membranes. Most eukaryotic cell membranes contain high concentrations of cholesterol, which is composed of

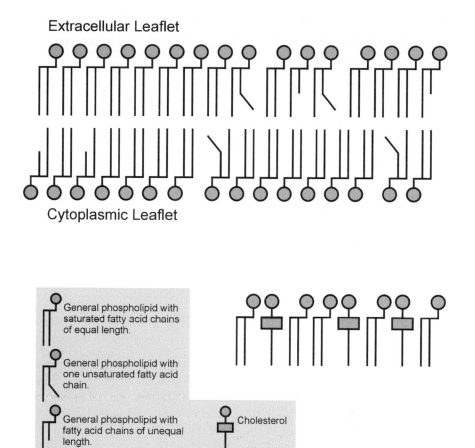

Figure 1-1 Organization of the lipid bilayer.
In a hydrophilic environment, phospholipids are arranged as a bilayer. In the extracellular layer, the polar groups of the phospholipids face the extracellular fluid, whereas the polar groups of the phospholipids forming the cytoplasmic leaflet face the watery environment of the cytoplasm. The nonpolar hydrocarbon chains of the lipids in both leaflets face the interior of the bilayer. Membrane phospholipids may have two identical hydrocarbon chains, chains of unequal length, or one saturated and one unsaturated (presence of one or more double bonds between carbons) fatty acid chains. Cholesterol is also a principal component of biologic membranes.

a central rigid structure of steroid rings, a polar hydroxyl group, and a nonpolar hydrocarbon chain. The rigid steroid rings restrict the movement of the fatty acid chains of adjoining lipids. However, at high concentration in the membrane, cholesterol may increase membrane fluidity by preventing close packing of lipids (Fig. 1-1).

Permeability to water and small water-soluble molecules increases in parallel with membrane fluidity.

MAJOR CONSTITUENTS OF THE LIPID BILAYER

Phospholipids
   Phosphatidyl choline
   Phosphatidyl serine
   Phosphatidyl ethanolamine
   Sphingomyelin
Glycolipids
   Gangliosides
   Galactocerebrosides
Cholesterol

Cell membranes are asymmetric in lipid constitution.

Until now, this chapter has been considering lipid bilayers as forming spontaneously in a water environment. In that case, components might be expected to be randomly and uniformly distributed between the two halves of the lipid bilayer. *Cell membranes, however, are actively constructed in the endoplasmic reticulum* of the cell with the aid of enzymes (e.g., flipases). (See Chap. 2.) In the endoplasmic reticulum, uncharged phospholipids such as phosphatidyl choline and sphingomyelin are directed to the half of the bilayer which will form the extracellular face. *Note that this will be the inside layer of the membrane-bound vesicles which leave the endoplasmic reticulum* (Fig. 1-2). Phosphatidyl ethanolamine and negatively charged phosphatidyl serine are primarily directed to the portion of the bilayer which will face the cytoplasm. The resulting concentration of negative charge on the cytoplasmic surface of the cell membrane may play a role in directing enzymes to this surface.

Glycolipids are also unevenly distributed in the bilayers of the various cell membranes (Fig. 1-2). The polar heads of glycolipids contain one or more sugar residues, while some also contain negatively charged sialic acid groups. *Glycolipids are exclusively distributed to the noncytoplasmic half of the lipid bilayer* where they may play a role in cell-cell and cell-matrix interactions. They are particularly abundant in the extracellular half of the cell membrane of the myelin sheath surrounding nerve fibers.

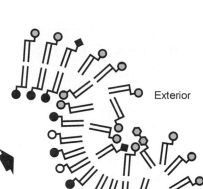

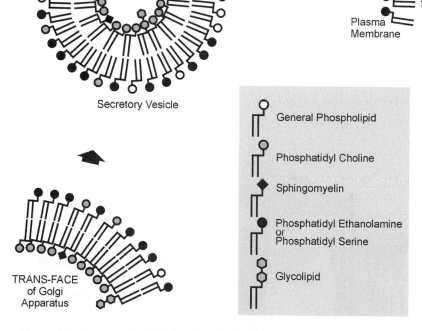

Figure 1-2 Asymmetric distribution of specific phospholipids to the two leaflets of the lipid bilayer.

Specific phospholipids are enzymatically distributed to membrane leaflets in the rough endoplasmic reticulum (RER, not shown). The distribution of these lipids is maintained when vesicles bud from the *trans*-face of the Golgi apparatus. Note that when these vesicles fuse with the plasma membrane, the leaflet facing the interior of the vesicle becomes part of the extracellular leaflet, whereas the vesicular leaflet facing the cytoplasm becomes part of the cytoplasmic leaflet of the plasma membrane.

MEMBRANE PROTEINS

The fluid mosaic model, in its most simplistic form, describes a flotilla of protein "icebergs" floating in a sea of lipid (Fig. 1-3). Real biologic membranes are only slightly more complicated with respect to the association of proteins with the lipid bilayer.

Proteins associated with the lipid bilayer are either stuck in it (integral) or stuck on it (peripheral).

Integral membrane proteins associate with the nonpolar core of the lipid bilayer either directly or indirectly. Integral membrane proteins include proteins which span the lipid bilayer one or more times and proteins which are anchored to the lipid bilayer but do not pass through it (Fig. 1-4).

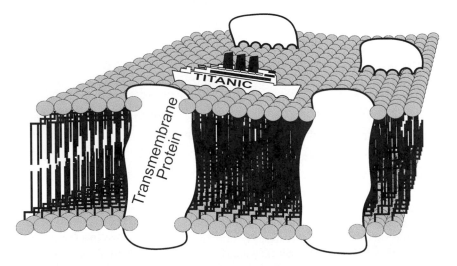

Figure 1-3 The protein "iceberg" model of cellular membranes.
This model specifically displays transmembrane proteins as icebergs floating in a sea of lipids. Other integral and peripheral membrane proteins are not shown.

Multiple-Pass
Transmembrane Protein

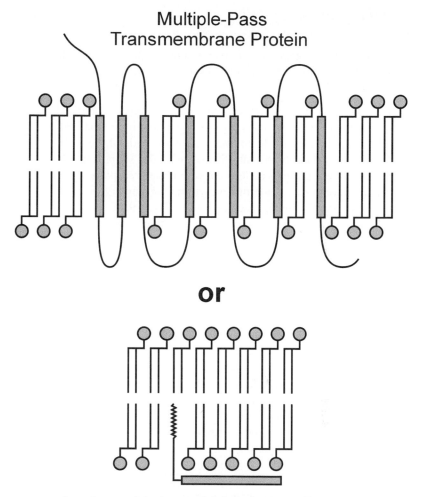

or

Anchored Integral Membrane Protein

Figure 1-4 Integral membrane proteins.
Integral membrane proteins include the transmembrane proteins. These proteins contain one or more hydrophobic portions that are inserted in the hydrophobic core of the lipid bilayer while the hydrophilic portions of the protein extend into the extracellular and cytoplasmic environments. The one shown in the diagram is a seven-pass transmembrane protein typical of a particular class of cell receptor proteins (upper figure). Anchored proteins are another class of integral membrane proteins. Rather than extending across the entire membrane, they are anchored by a hydrophobic tail into one leaflet of the bilayer.

Integral membrane proteins

Transmembrane proteins contain one or more sequences of predominantly hydrophobic amino acids embedded in the lipid bilayer.

A *single-pass* transmembrane protein contains a single hydrophobic sequence of about 25 amino acids in an α-helical configuration which is embedded in the nonpolar core of the bilayer. The polar ends of the amino acid chain extend into the cytoplasm and into the extracellular soup. Examples of single-pass transmembrane proteins include enzyme-linked receptors such as the receptor tyrosine kinases (e.g., epidermal growth factor, nerve growth factor, insulin, and natriuretic peptide family receptors). (See Chap. 9.) Glycophorin, the major transmembrane protein of erythrocyte plasma membranes, is also a single-pass protein.

Multiple-pass transmembrane proteins, as their name would suggest, contain more than one α-helical hydrophobic amino acid sequence. Each hydrophobic sequence is embedded in the lipid bilayer and linked by extracellular and intracellular sequences of polar amino acids. Quite frequently, these hydrophilic sequences serve as sites for ligand binding or phosphorylation. (Addition of a phosphate group usually activates these proteins.) *G-protein-linked receptors* are examples of multiple-pass transmembrane proteins. These proteins typically have seven transmembrane domains (stitches) and include the receptors for many neurotransmitters including acetylcholine, the catecholamines, and various neuropeptides. Other examples of seven-pass transmembrane proteins include cell-surface receptors for the HIV virus. Multiple-pass transmembrane proteins also include the large transporters and channel-forming proteins such as *porin*, which has 12 transmembrane domains.

Some integral membrane proteins are anchored only indirectly to the lipid bilayer.

Another class of integral membrane proteins is associated with the lipid bilayer, not by internal nonpolar amino acid sequences, but by a *hydrophobic anchor* inserted in the bilayer. Some members of this group are covalently linked to a fatty acid chain embedded in the bilayer (Fig. 1-4). Examples include the Src and Ras proteins which participate in the activation of mitogen-associated protein (MAP) kinases. (See Chap. 9.) Other members of this group are linked by sugar residues to phosphatidyl inositol, a minor constituent of the lipid bilayer. These proteins are generally refered to as *glycophosphatidyl inositol (GPI)-linked* proteins (Fig. 1-5). *Phosphatidyl inositol is asymmetrically distributed in the lipid*

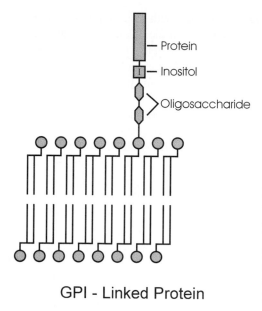

GPI - Linked Protein

Figure 1-5 GPI-linked proteins.
Another type of anchored membrane protein is linked to phosphatidyl inositol, a minor component distributed specifically to the extracellular leaflet of the plasma membrane.

bilayer. Therefore, GPI-linked proteins are found only on the external surface of the cell membrane. One example of this type of protein is the *folate receptor*.

Integral membrane proteins may be removed from cell membranes only by detergents.

Remember that integral membrane proteins are either embedded in, or anchored to, the hydrophobic core of the lipid bilayer. They can be freed from the membrane only by agents which disrupt the lipid bilayer. Amphipathic detergents possess both a polar and a nonpolar region. The nonpolar region is able to invade the bilayer and interact with the nonpolar region of the integral protein. This leaves the polar region of the detergent free to interact with water; thus the protein-detergent complex becomes water soluble and leaves the bilayer.

Peripheral membrane proteins

Peripheral membrane proteins interact with other proteins on the membrane surfaces by ionic forces and may be removed by altering the ionic content or pH of their environment.

Peripheral membrane proteins may associate with either integral membrane proteins or other peripheral membrane proteins by ionic linkage. For example, the cytoskeletal protein spectrin, found in red blood cells, is linked ionically to ankyrin, another peripheral membrane protein which is, in turn, ionically linked to the cytoplasmic portion of the integral transmembrane protein, band 3 protein (Fig. 1-6).

Subsequent chapters discuss how membranes form the boundaries of cytoplasmic organelles (Chap. 2), how proteins are inserted through or into membranes (Chap. 5), and how cellular membranes are recycled (Chap. 6).

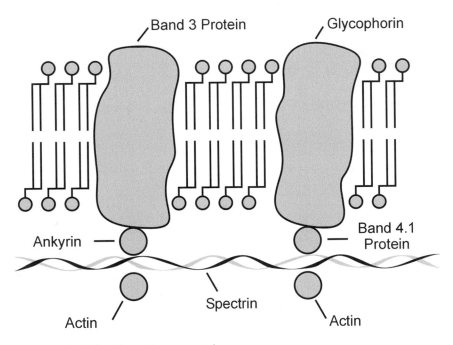

Figure 1-6 Peripheral membrane proteins.
Peripheral membrane proteins (e.g., ankyrin and band 4.1 protein) are ionically linked either directly to integral membrane proteins, or indirectly (spectrin) via other peripheral proteins.

CYTOPLASM

·

- **Ribosomes**
- **Endoplasmic Reticulum**
- **Golgi Apparatus**
- **Lysosomes**
- **Secretory Vesicles**
- **Mitochondria**
- **Peroxisomes or Microbodies**
- **Cytosol**

· · · · · · · · · · · ·

The cytoplasm contains a variety of *organelles* embedded in an aqueous matrix. The organelles are described in this chapter, whereas the function of organelles in trafficking and motility is discussed in subsequent chapters. The components of the cytoskeleton including microfilaments, microtubules, and intermediate filament proteins are discussed in Chap. 3.

RIBOSOMES

Ribosomes are particles composed of RNA and protein, and are essential for protein synthesis. Ribosomes *translate* the coded message transcribed from the genes in the form of messenger RNA (mRNA).

Ribosomes are found as individual structures or in groups (polyribosomes or polysomes) in two different arrangements: (1) attached to membranous cisternae (endoplasmic reticulum) and (2) freely suspended in the cytoplasm. These two arrangements differ in the types of proteins being synthesized.

Unattached polyribosomes are involved in the synthesis of intracellular proteins, whereas polysomes attached to the cisternae of the endoplasmic reticulum are involved in protein synthesis for export from the cell or for distribution to lysosomes (organelles specialized for degradative processes).

Polysomes consist of numerous ribosomes attached to the same strand of messenger RNA (see Fig. 5-3). Ribosomes move along the chain from $5' \rightarrow 3'$, synthesizing protein as they travel. The advantage of the polyribosome complex is that more than one ribosome is attached to the same mRNA at the same time, rapidly generating polypeptide chains which are exactly alike.

The ribosomal subunits and associated proteins are synthesized in the nucleolus found within the nucleus.

The subunits are assembled in the cytoplasm. Prokaryotic and eukaryotic ribosomes differ in their arrangement, although both consist of small and large subunits.

Ribosomes are typically defined by their sedimentation coefficients. *Eukaryotic ribosomes are defined as 80S with a large (60S) and a small (40S) subunit.* Since these are sedimentation coefficients, they are not additive. The 60S subunit contains 28S, 5S, and 5.8S ribosomal RNA (rRNA) while the 40S subunit contains an 18S rRNA. A variety of proteins make up the remainder of the large and small subunits. The ribosomal subunits self-assemble in the presence of mRNA. A more detailed discussion of ribosomal subunits is provided in Chap. 4, "Nucleus."

The ribosome contains three RNA-binding sites: one for messenger RNA on which the ribosome sits and the other two for transfer RNAs.

Ribosomes move along the messenger RNA in a stepwise fashion. The growing peptide chain extends from the *peptidyl-tRNA-binding (P) site*. The other transfer RNA binding region is known as the *aminoacyl-tRNA-binding (A) site* and holds the incoming amino acid which is being added to the growing peptide chain. Chapter 5, "Protein Synthesis," provides a more complete explanation of ribosomal protein synthesis.

ENDOPLASMIC RETICULUM

The endoplasmic reticulum (ER) is a continuous cytomembrane system of cisternae which links the synthetic machinery of the cell to the outside environment. The ER consists of two parts which are described by the absence (smooth, SER) or presence (rough, RER) of ribosomes. This is the cytoplasmic analogy to "Sometimes you feel like a nut, sometimes you don't."

The *ER* is an *enclosed cisternal system* which defines the lumen and separates it from the remainder of the cytoplasmic glop (matrix), also known as cytosol, in which organelles exist. The rough endoplasmic reticulum (RER) and smooth endoplasmic reticulum (SER) are interrelated, but have separate functions within the cell. Although the presence of extensive RER including the associated ribosomes can affect the staining quality of cells at the light-microscopic level (e.g., Nissl staining in multipolar neurons or the exaggerated polarity of pancreatic acinar cells), the ER can be visualized only by electron microscopy. It can also be isolated by differential centrifugation using a sucrose gradient, a method which has been used to separate many of the organelles of the cell. The oriented stacks of flattened cisternae which constitute the endoplasmic reticulum are shown in Fig. 2-1.

RER and its attached ribosomes are the site of protein synthesis, whereas the SER functions in detoxification and lipid synthesis.

RER is abundant in cells which are factories for exportable protein, such as the plasma cell which makes antibodies. SER is abundant in steroid-producing cells and increases in volume in the liver of animals after treatment with agents such as barbiturates.

The ER membrane separates the cisternal system from the cytosolic compartment.

Exportable and transmembrane proteins must be translocated into the cisternal space of the ER to reach their final destination. These proteins are synthe-

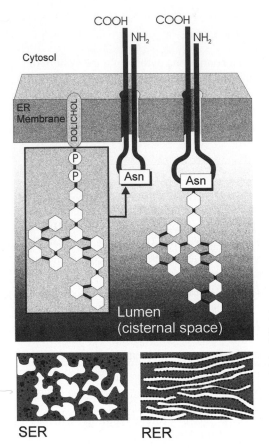

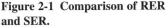

Figure 2-1 Comparison of RER and SER.
The top figure illustrates the process of N-linked glycosylation which occurs through the action of dolichol, a lipid carrier within the RER. The lower figures compare the EM appearance of SER and RER.

sized by polyribosomes and translocated across the ER membrane by a process known as the *signal hypothesis* (see Chap. 5, "Protein Synthesis"). Proteins synthesized on free ribosomes are not translocated into the cisternal space and remain in the cytoplasm or are targeted to mitochondria or the nucleus.

> The SER synthesizes phospholipids and cholesterol, which are the lipids necessary for turnover of cell membranes—plasma membrane as well as the membranes of organelles (see Chap. 1, "Membranes").

   In comparison to protein synthesis, *lipid synthesis occurs in association with the cytosolic leaflet of the ER; translocation to the lumenal leaflet occurs through the action of phospholipid translocator proteins, also known as flippases.*

> Transport of lipid and phospholipid to the cell membrane and organelles occurs by vesicular transport or by phospholipid transfer proteins.

Transport of lipid to the cell membrane and to most organelles occurs through vesicular transport. Phospholipid is transferred from the ER to mitochondria and peroxisomes by water-soluble phospholipid transfer (exchange) proteins.

> The polarity of organellar membranes is established in the ER.

Polarity of organellar membranes is established by *asymmetric lipid synthesis, protein insertion, and glycosylation.* Phospholipid translocators (flippases) are responsible for the asymmetry of lipid synthesis. Other organelles upstream are not abundant in flippases; therefore, the ER leaflet position of lipids remains essentially unchanged. Protein polarity is established at the time that the nascent peptide is inserted into the ER membrane. Glycosylation also contributes to the polarity of organelle membranes, and this process is initiated in the RER.

> There are two types of glycosylation (N-linked and O-linked) which differ in the location of the oligosaccharide on the protein and the location of the process within the cell as well as in the synthetic mechanism itself.

Sugar residues are connected to protein through either a N-linkage or an O-linkage. N-linked refers to the connection of oligosaccharides to the amino group of asparagine (NH_2), whereas O-linked refers to connection through the hydroxyl group of threonine, serine, or hydroxylysine (OH^-).

> N-linked glycosylation begins in the RER by a mechanism involving a dolichol carrier.

N-linkage is initiated through the addition of a 14-sugar, branched oligosaccharide molecule to the NH_2 of asparagine when this amino acid occurs in a specific tripeptide recognition sequence. The transfer process is known as *en bloc transfer* because an identical oligosaccharide is added to each protein to be glycosylated. The transfer is accomplished by the action of an oligosaccharide transferase and the membrane-bound lipid carrier, *dolichol*.

The branched oligosaccharide is trimmed and partially processed in the RER. The remainder of oligosaccharide processing occurs in the Golgi apparatus. This includes further trimming and processing of the branched oligosaccharide as well as the O-linked glycosylation process.

GOLGI APPARATUS

The Golgi apparatus has several major functions. It is (1) the sorting station of the cell, (2) the location of posttranslational modifications including glycosylation (which is initiated in the RER) and sulfation, and (3) the site for the assembly of proteoglycans.

The Golgi apparatus is a prominent feature of cells dedicated to secretion. *Glycoproteins and glycolipids are the primary products which pass through the Golgi apparatus.* The Golgi apparatus sorts products into different pathways. Vesicles are targeted from the Golgi to the apical or basolateral membranes of the cell. Transport vesicles carry partially glycosylated products from the RER which enter the flattened stacks of the Golgi apparatus. Exit occurs either into a secretory pathway or a lysosomal storage pathway.

The Golgi apparatus has a distinct functional and structural polarity.

The Golgi apparatus (Fig. 2-2) consists of a stack of flattened disks or cisternae with a distinct polarity. Each region is called a *network, face, or cisterna.* The entrance face is called the *cis-Golgi network* (CGN). This region receives transfer vesicles (coated with a protein called *coatomer*) which bud from the transitional elements of the RER and is the major site of phosphorylation of oligosaccharides located on lysosomal proteins. The next region is collectively known as the *Golgi stack* and consists of subdivisions or compartments: the *cis*, medial, and *trans* cisternae. The exit face is called the *trans-Golgi network* (TGN).

N-linked glycosylation is completed in the Golgi apparatus.

N-linked glycosylation occurs through a series of processing steps involving the activity of specific enzymes classified as glycosyl transferases.

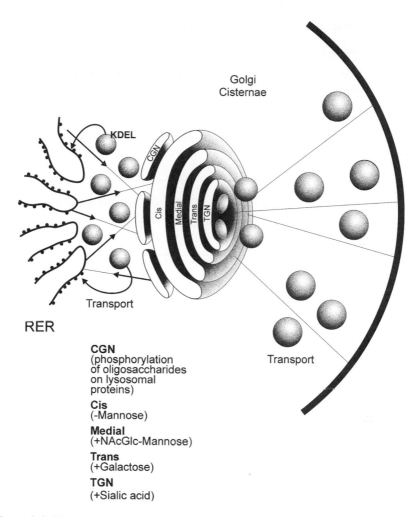

Golgi
Cisternae

KDEL

Transport

RER

CGN
(phosphorylation
of oligosaccharides
on lysosomal
proteins)

Cis
(-Mannose)

Medial
(+NAcGlc-Mannose)

Trans
(+Galactose)

TGN
(+Sialic acid)

Transport

Figure 2-2 The compartmentalization of Golgi function.
This *cis*-Golgi network (CGN), *trans*-Golgi network (TGN), and the Golgi cisternae (cis, medial, and trans) contain enzymes involved in specific Golgi functions. The direction of transport is from RER → CGN → *cis* → medial → *trans*-cisternae → TGN. RER proteins contain a KDEL (Lys-Asp-Glu-Leu) signal resulting in their return to the RER.

O-linked glycosylation does not occur by an *en bloc* method.

O-linked glycosylation is catalyzed by the action of specific Golgi-membrane-bound oligosaccharide transferases. These enzymes are associated with different regions of the Golgi stacks.

The cisternae of the Golgi are shown in Fig. 2-2.

The *cis*, medial, and *trans* cisternae are the location of mannose residue removal and addition of galactose and *N*-acetyl-glucosamine.

The compartments of the Golgi apparatus are discussed in more detail in Chap. 6, "Intracellular Trafficking." (See Fig. 6-2.)

In comparison to the *cis*-Golgi network (CGN) which is the docking area for transport vesicles from the RER, the exit face is known as the *trans*-Golgi network (TGN). The TGN is the major sorting area of the secretory cell.

The TGN sorts vesicles to the apical or basolateral membranes (domains). In some cases errors in vesicular transport to the correct domain are corrected through endosomes and the process of endocytosis (see Chap. 6).

The Golgi apparatus is the site of proteoglycan assembly. The protein portion is synthesized in the RER, but the intensive glycosylation and assembly are Golgi events.

Proteoglycans are key extracellular matrix molecules which will be discussed in Chaps. 13 and 14. These molecules consist of *glycosaminoglycans* (GAGs, unbranched sugar chains consisting of repeating units of specific sugar molecules) and core and link proteins. The core and link proteins are synthesized in the RER and reach the Golgi apparatus by transport vesicles. GAGs are synthesized in the Golgi apparatus and added to the core proteins to form *proteoglycan subunits*. These subunits are joined by link proteins to a central GAG, often *hyaluronan,* to form the *aggregate.*

Sulfation occurs in the Golgi apparatus.

Molecules such as the proteoglycans are heavily sulfated. This process occurs in the TGN for the proteoglycans and for other proteins which are sulfated at their tyrosine residues.

LYSOSOMES

Lysosomes are membranous organelles which contain acid hydrolases for the controlled degradation of macromolecules.

Lysosomal enzymes are hydrolases which function at an acidic pH, an environment maintained within the lysosome by a H^+ pump. The enzymes are very specific; in the absence of a specific enzyme, the undigested substrate builds up in the lysosome. Lysosomal storage diseases result from the absence of even a single acid hydrolase (see table below). In a rare disease known as *I-cell disease* there is a genetic defect which results in the absence of all the acid hydrolases from the lysosomes. The lysosomal enzymes are missorted to the secretory pathway because of the absence of the signal required for entrance into the lysosome. Only cell types lacking a secondary mechanism for hydrolase transport to lysosomes are affected.

CHARACTERISTICS OF LYSOSOMAL STORAGE DISEASES

LYSOSOMAL STORAGE DIEASE	ENZYME DEFICIT	CELLULAR SITE/ ACCUMULATION		ORGAN(S) MOST AFFECTED
Tay-Sachs	β-*N*-hexosaminidase-A	Neurons	Glycolipid	CNS
Gaucher	β-D-glycosidase	Macrophages	Glycolipid	Spleen, liver
Hurler	α-L-iduronidase	Fibroblasts, chondroblasts, osteoblasts	Dermatan sulfate	Skeletal system
Niemann-Pick	Sphingomyelinase	Oligodendro-cytes, fibro-blasts	Sphingo-myelin	CNS
Inclusion (I)-cell	*N*-acetylglucosamine phosphotransferase	Fibroblasts, macrophages	Glyco-proteins, glycolipids	Nervous and skeletal systems, (liver unaffected)

Lysosomes function in the endocytic pathway (incoming intracellular traffic route) although there are three pathways which lead to lysosomes: (1) *endocytosis*, the process by which small molecules are taken up by the cell and directed to vesicles called *endosomes*, which eventually form lysosomes; (2) *autophagy* in which worn-out organelles are broken down by lysosomal activity; and (3) *phagocytosis* in which worn-out organelles are broken down by cells specialized for the uptake of bacteria, viruses, and large particles.

SECRETORY VESICLES

The secretory vesicle or granule is a storage and concentration site for synthesized secretory product. However, it is also the site of proteolytic processing for some proteins.

Secretory vesicles bud off from the TGN and are coated with a protein called *clathrin*, identical with the coating of vesicles containing materials entering the cell (endocytosis). The clathrin is removed during the maturation of the vesicle as the secretory product within the granule is concentrated. Proteolytic processing often occurs in the secretory vesicles resulting in removal of the pro-portion of a protein molecule. Peptides such as insulin are synthesized as preproinsulin. The pre-segment includes the signal peptide, which is removed on formation of the nascent peptide in the RER. The pro-portion is separated from insulin in the secretory vesicles through the action of a H^+ pump, which acidifies the vesicle contents. Information on protein processing has been obtained by use of chloroquine, an antimalarial drug, which blocks the acidification process. These processes are discussed in greater detail in Chap. 5, "Protein Synthesis" and Chap. 6 "Intracellular Trafficking."

Secretory vesicles fuse with the cell membrane, their contents are released, and the membrane components are recycled by endocytosis.

Secretory vesicles fuse with the cell membrane in a two-phase process, which includes docking and fusion. Following fusion, the reverse process of endocytosis is used to transport membrane to the endosomal compartment. Addition of membrane to the plasma membrane is balanced by endocytosis, which removes membrane by way of endosomes back to the TGN.

MITOCHONDRIA

Mitochondria generate usable energy for the cell in the form of ATP by establishing a transmembrane electrochemical proton (H^+) gradient.

A mitochondrion has two membranes: an inner and an outer, which separate 2 compartments, an intermembranous space (between the two membranes) and the matrix which fills the space surrounded by the inner membrane. The gradient is established through proton translocating activity in the inner membrane associated with the respiratory chain and mitochondrial ATP synthase (Fig. 2-3).

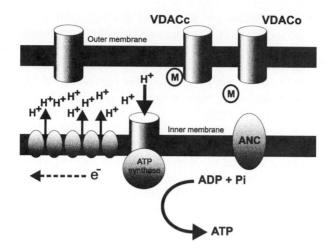

Figure 2-3 The chemiosmotic hypothesis.

There is a specific localization of molecules within the mitochondrial membranes and compartments.

MITOCHONDRIAL REGION	CONTENT	FUNCTION
Outer membrane	Porin VDAD (voltage-dependent anion channel)	Aqueous channel for molecules under 10 kDa
Inner membrane	Respiratory chain, ATP synthesis, transport proteins, cardiolipin	Proteins for generation of energy and export and import; cardiolipin creates impermeability to small ions
Matrix	Citric acid cycle enzymes, mitochondrial DNA, ribosomes	Production of CO_2 and NADH [the primary source of electrons for the electron transport (respiratory) chain]; mitochondrial genome and machinery for expression of mitochondrial genes
Intermembrane space	Enzymes	Use ATP for nucleotide phosphorylation

The mitochondrion breaks down pyruvate and fatty acids to acetyl coenzyme A (acetyl CoA) which enters the Krebs cycle leading to generation of high-energy electrons passing through the respiratory chain in the inner mitochondrial membrane. The function of the respiratory chain is to transport the electrons to oxygen, leading to oxidative phosphorylation; the conversion of ADP + P_i → ATP. This system is based on the maintenance of a chemiosmotic coupling system which is shown in Fig. 2-3.

As electrons released by oxidation of substrate in the matrix flow down the respiratory chain, hydrogen ions (H^+) are pumped into the intermembrane space. Because the inner mitochondrial membrane is impermeable to hydrogen ions, an osmotic and electrochemical gradient develops. The gradient forces hydrogen ions through the ATP synthase complex, directing the synthesis of ATP from ADP and P_i and storing energy in the phosphate bond. This functions like water turning a waterwheel with the H^+ gradient driving the ATP synthase. The adenosine nucleotide cotransporter (ANC) allows ATP to be transported out of the matrix compartment and ADP to be transported in. Also depicted are the voltage-dependent anion channels (VDAC = porin) in the outer mitochondrial membrane. Porin, or VDAC, is regulated by a moderator protein. In the presence of the moderator (M) the anion channel is closed ($VDAC_C$) when the moderator is released the channel is open ($VDAC_O$). When the voltage-dependent anion channels are open, the ionic gradient is at zero. When the voltage-dependent anion channels are closed either a positive or negative voltage exists.

PEROXISOMES OR MICROBODIES

Peroxisomes are single membrane-bound organelles which carry out oxidative reactions producing hydrogen peroxide.

Peroxisomes contain high concentrations of catalase, the enzyme which utilizes hydrogen peroxide as a substrate to oxidize other molecules such as ethanol. In the liver the predominant cell type, the hepatocyte, contains numerous peroxisomes which convert ethanol to acetaldehyde. Peroxisomes also have a major function in breakdown of fatty acid molecules.

CYTOSOL

The cytosol is that part of the cell which contains the organelles, the cytoskeleton, and a variety of other molecules: water, ions, proteins, glucose, adenosine triphosphate, amino acids, and nucleotides.

The cytosol including the cytoskeletal elements constitutes approximately one-half of the cellular volume. It is an intracellular matrix which gives shape and form to the cell and is not randomly arranged. The cortical cytoplasm just under the plasma membrane has a distinct arrangement of cytoskeletal elements which excludes organelles and maintains the stiff gel form of the cell.

When the cell moves as required during certain cellular activities, this gel-like state is liquefied through the action of specific proteins which bind and allow rapid polymerization of some cytoskeletal components. The action of these proteins is described in Chap. 3, which reviews the structure and function of the cytoskeleton.

CYTOSKELETON

•

- **Microfilaments (Actin)**
- **Spectrin Cytoskeleton**
- **Microtubules (Tubulin)**
- **Intermediate Filaments**

• • • • • • • • • • • •

The cellular biomechanical framework (cytoskeleton) consists of three specific types of proteins alternating between soluble subunits and organized fibrillar or tubular structures: microfilaments, intermediate filaments, and microtubules. A fourth class includes the spectrins and related proteins. Components of the cytoskeleton form the scaffolding or framework which defines cell shape and resiliency. The cytoskeletal framework also permits the distribution of force between adjacent cells and allows the anchoring of cells to each other and the underlying basal lamina. Finally, the cytoskeleton is intimately involved in cell motility, cell division, and the movement of intracellular organelles.

MICROFILAMENTS (ACTIN)

Networks of actin microfilaments play a major role in maintaining cell shape and the integrity of fragile cellular projections such as microvilli. Actin microfilaments support the cell membrane by binding to a number of specific transmembrane proteins either directly or indirectly. A region of high actin concentration immediately beneath the plasma membrane, the *actin cortex*, plays an important role in motility and cell division. Actin is concentrated in other regions of the cell, including the core of microvilli and the contractile ring which separates daughter cells during mitosis.

> Actin exists in the cytoplasm in a dynamic equilibrium between the monomeric form (G-actin) and the polymerized, fibrillar form (F-actin).

Although the types of actin found in muscle (α-actin) and nonmuscle cells (β- and γ-actin) vary slightly, they share a highly conserved amino acid sequence and behave in a similar fashion biochemically. Actin exists both as single actin molecules (G-actin) (Fig. 3-1) and as long chains (F-actin) formed by the self-assembly of G-actin monomers.

> G-actin is polarized with plus and minus ends. When G-actin polymerizes, it forms a filament which also has plus and minus ends.

The rate of addition of new G-actin to actin filaments occurs more rapidly and at a lower G-actin concentration at the plus end than at the minus end. Very low concentrations of G-actin favor the disasssembly of actin filaments, whereas intermediate concentrations favor a dynamic equilibrium between net subtraction at the minus end and net addition at the plus end (*treadmilling*) (Fig. 3-2). Monomers are added continuously at the plus end, and equal amounts are removed continuously at both the plus and minus ends resulting in zero net growth of the filament. Higher concentrations of G-actin favor net addition at both ends and thus growth of the actin filament.

> Polymerization of actin requires ATP, Mg^{2+}, and K^+.

ATP is tightly bound to G-actin. Soon after a G-actin monomer is added to the actin filament, its ATP is hydrolyzed to ADP + P_i (inorganic phosphate). The ADP remains trapped within the polymerized actin and increases the likelihood of depolymerization. Depolymerized G-actin slowly exchanges ADP for ATP and is then able to polymerize again.

Figure 3-1 Polarized actin monomer.

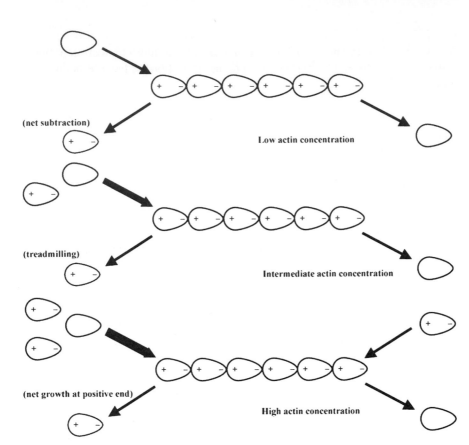

Figure 3-2 Polymerization of actin at different concentrations of monomer.
Width of arrows indicates relative amount of addition or subtraction. Treadmilling occurs
when addition at plus end equals sum of subtraction at both plus and minus ends.

Small peptides regulate the polymerization of actin.

The normal intracellular concentration of actin is high enough that the
majority of actin should be in the form of F-actin microfilaments. However, this
does not usually occur. A variety of small peptides (see table below) bind to
G-actin and moderate its ability to form polymers, whereas others stabilize or
destabilize the actin filaments. The G-actin concentration within the cell is quite
high (about 200 μM), yet the actin-binding peptides (ABPs) actually keep the
free monomer concentration to about 0.2 μM. This mechanism provides a large

pool of accessible G-actin and helps to regulate the polymerization of actin in response to environmental stimuli. The diversity of function of actin filaments in different regions of the cell is generated by the actin-binding proteins and not by the structure of actin.

ACTIN-BINDING PEPTIDES

ACTIN-BINDING PROTEIN	FUNCTION
Spectrin and dystrophin	Bind cortical cytoskeleton to plasma membrane
Villin and Fimbrin	Cross-link densely in microvilli
Fimbrin	Cross-links in microspikes and lamellipodia
Calmodulin and myosin I	Cross-link actin to plasma membrane in microvilli
α-Actinin	Cross-links stress fibers and connects actin to protein–plasma membrane complexes
Filamin	Cross-links actin at wide angles to form screen-like gels
Gelsolin	Destabilizes F-actin and caps it to prevent renucleation
Myosin V	Facilitates plus-end–directed vesicle movement
Myosin II	Generates contraction or tension
Tropomyosin	Stabilizes F-actin, prevents filamin-binding, increases myosin II binding
Profilin	Binds actin monomers
Cap Z	Caps F-actin plus end and provides long-term stability

Actin microfilaments (F-actin) consist of a double helical (coiled-coil) chain of G-actin subunits. (Fig. 3-3).

Figure 3-3 Coiled-coil dimer of actin α-helices.

SPECTRIN CYTOSKELETON

Spectrin filaments form a strong, yet deformable, lattice work on the cytoplasmic surface of the plasma membrane. The spectrin cytoskeleton stabilizes and maintains the shape of the plasma membrane and connects the plasma membrane to other cytoskeletal components, especially actin. First discovered in association with the erythrocyte plasma membrane, members of the spectrin family of proteins, as well as other associated transmembrane and peripheral proteins, are now known to be present in most animal cells.

Spectrin is a fibrous protein consisting of α and β subunits arrayed as an antiparallel dimer.

In the erythrocyte membrane, spectrin (spectrin I) forms tetramers by linking two dimers end-to-end. Each end of the *tetramer* contains an actin-binding site. Spectrin binds short (10 to 20 monomers) actin filaments associated with tropomyosin.

Spectrin attaches indirectly to the cytoplasmic (amino) portions of transmembrane proteins.

In the erythrocyte membrane, spectrin is linked to *band 3 protein*, a transmembrane anion transporter, through an intermediary protein, *ankyrin* (Fig. 3-4). Spectrin is also linked to transmembrane proteins of the *glycophorin* family through *band 4.1 protein* which also greatly increases the affinity of actin for spectrin.

The spectrin cytoskeleton permits erythrocytes to deform when passing through narrow capillaries and allows them to return to their normal, biconcave shape in less restrictive confines.

Mutations in the various genes coding for proteins contributing to the spectrin cytoskeleton can result in defects in the erythrocyte membrane leading to

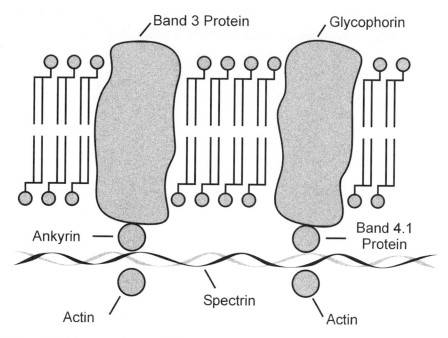

Figure 3-4 The spectrin cytoskeleton.
Spectrin filaments attach indirectly to the transmembrane proteins band 3 protein and glycophorin through the peripheral membrane proteins ankyrin and band 4.1 protein, respectively.

increased fragility and reduced deformability. Examples include various types of hereditary hemolytic anemia and irreversibly sickled cells in sickle cell disease (anemia).

MICROTUBULES (TUBULIN)

Microtubules provide another structural foundation for the transport of organelles and vesicles within the cytoplasm. They also form the machinery for the motion of cilia and flagella as well as the spindle apparatus for the separation of chromosomes during meiosis and mitosis. Theses functions are discussed in detail in Chap. 7 "Cell Motility," and Chap. 10, "Cell Cycle." Microtubules share many functional and organizational characteristics with the actin monomer-filament system.

Microtubules are composed of pairs (dimers) of α- and β-tubulin subunits.

The α/β subunits of tubulin are assembled into long, linear arrays or *protofil-aments* (Fig. 3-5). Thirteen protofilaments are arranged around the circumference of an imaginary column to form the walls of a hollow tube 25 nm in diameter.

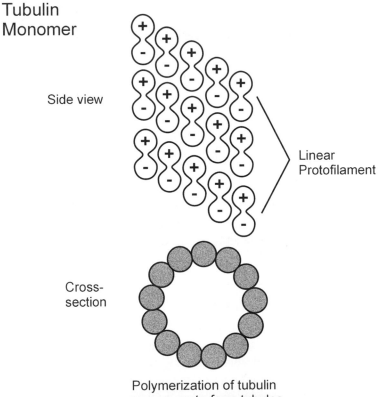

Figure 3-5 Tubulin monomers, protofilaments and microtubules.
Tubulin monomers (upper figure) consisting of α and β subunits are arranged in linear arrays to form protofilaments (middle). Thirteen protofilaments form a microtubule (lower figure).

Since tubulin subunits are polar, they form microtubules with fast-growing (plus) ends and slow-growing (minus) ends.

In similar fashion to actin microfilaments, microtubules add subunits more rapidly at their plus ends than at their minus ends. For this reason, their plus ends appear to elongate away from nucleation sites such as the basal bodies and centrosomes. Most, but not all, *microtubule organizing centers (MTOC)* consist of centrioles (Fig. 3-6) or centriole-like elements (basal bodies), singly or in pairs, surrounded by a matrix of specific minor tubulins (e.g., γ-*tubulin*) and associated proteins. It is the matrix which actually performs the nucleation of microtubules (see Chap. 10 for the organization of actin filaments of the mitotic spindle).

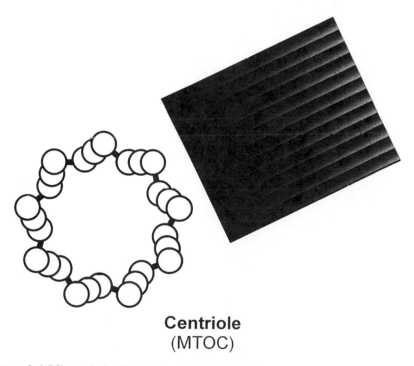

Centriole
(MTOC)

Figure 3-6 Microtubule organizing center (MTOC).
MTOCs may occur singly (basal bodies) or in pairs (centrioles) as shown in this figure. The second centriole (side view) is oriented perpendicularly to the first centriole (end view). Centrioles form the basic structure of the centrosome, which organizes the mitotic spindle and plays an important role in cell division. Basal bodies are the organizing nucleus for the axoneme of cilia and flagella.

> Individual microtubules exist in a continuous state of assembly and disassembly (*dynamic instability*).

Both *in vitro* and *in vivo* microtubules demonstrate either steady growth or rapid disassembly, partially dependent on the concentration of free tubulin dimers. Unlike actin, the concentration of tubulin in the cytoplasm is generally low. Microtubules do not readily form or persist unless protected at their ends by capping. Capping of microtubular minus ends at the centrosomes removes a source of subtraction from microtubules, whereas capping at the plus ends protects from disassembly. At certain concentrations of monomer, microtubules may also undergo treadmilling (see pages 27–28).

> Polymerization of microtubules requires binding of GTP.

Although a molecule of GTP is bound to each of the α- and β-tubulins of a dimer subunit, only the GTP bound to the β-tubulin is hydrolyzable. After addition of a new dimer to the growing microtubule, the β-GTP is hydrolyzed to GDP + P_i, making the polymer more prone to disassembly. In a rapidly growing microtubule, addition of subunits occurs more rapidly than hydrolysis of GTP. This results in a protective GTP cap, which prevents disassembly of the growing microtubule.

> Chemical modifications and the actions of other proteins stabilize microtubules.

The α subunits of tubulin dimers are slowly modified by acetylation and detyrosination. These modifications provide binding sites for *microtubule-associated proteins (MAPs)* which stabilize microtubules. For instance, MAPs stabilize microtubules by binding to their sides like a reinforcing strut or by "capping" their ends, preventing further elongation or dissociation of the microtubule. MAPs may also confer structural and functional specificity on microtubules in specific "compartmentalized" portions of certain cells such as neurons. One class of MAPs, the tau proteins, is associated with the long microtubules of neuronal axons, whereas MAP-2 is associated with microtubules in the dendrites and cell body but is absent from the axon. MAPs may also link microtubules to other cytoskeletal components. Additionally, MAPs include the microtubule "motors," specifically members of the kinesin and dynein superfamilies. These are discussed in more detail in the Chap. 8, "Cell Motility."

INTERMEDIATE FILAMENTS

Intermediate filaments, named for their intermediate size (8 to 10 nm) between actin and myosin (or actin and microtubules), are typically found in abundance in cells subject to mechanical stress. They provide tensile strength in cells such as neurons and muscle and strengthen epithelia at cell-cell (desmosome) and cell-basal lamina (hemidesmosome) junctions. (See Chap. 7.) In addition, a special class of intermediate filament proteins, the lamins, provides support for the inner surface of the nuclear envelope. The lamins' function is discussed in detail in Chap. 4, "Nucleus," and Chap. 10, "Cell Cycle: Mitosis and Meiosis."

Intermediate filaments consist of several cell and tissue specific groups, plus the lamins.

INTERMEDIATE FILAMENTS

INTERMEDIATE-FILAMENT PROTEIN	LOCATION	COMMENT
Keratins (acidic and basic)	Epithelia	Acidic + basic = coiled-coil dimer
Desmin	Muscle	Links adjacent myofibrils in skeletal and cardiac muscle
Glial fibrillary acidic protein (GFAP)	Astrocytes	Routinely used as an astrocyte marker; increase in activated astrocytes
Vimentin	Cells of mesenchymal origin	Also found in immature cells such as developing astrocytes
Peripherin	Neurons	
Lamins (A, B, and C)	Inner face of the nuclear envelope	Form a two-dimensional sheet-like lattice; phosphorylation leads to dissociation in mitosis; dephosphorylation leads to nuclear envelope reassembly at the end of the mitotic cycle
Neurofilaments	Neurons	NF-L, NF-M, and NF-H in mature nervous systems

Intermediate filaments are composed of long, fibrous proteins.

Unlike actin filaments and microtubules which are both formed from globular subunits, intermediate filaments are formed by polymerization of long, rodlike proteins. The rodlike portion of the molecule consists of repeated amino acid segments (heptads) which facilitate the formation of coiled-coil α-helix dimers of parallel intermediate filament proteins. The carboxy and amino terminals of the proteins participate in links to other cytoskeletal elements. Two dimers in antiparallel array form a staggered *tetramer,* which is nonpolar, that is, the same at either end (Fig. 3-7). Tetramers bind end-to-end and pack together as eight strands to form a helical filament. *ATP and GTP are not required for filament polymerization.* Intermediate filaments form spontaneously, resulting in a low cytoplasmic concentration of soluble monomers.

Phosphorylation of the N-terminal of intermediate filaments may induce dissociation of the polymers.

Factors regulating the assembly of intermediate filaments are not well known. However, there is evidence that *phosphorylation* of the N-terminals of filaments may induce dissociation. For example, phosphorylation of nuclear lamins leads to their dissolution and the breakdown of the nuclear envelope. Subsequently, lamins are dephosphorylated prior to reassembly of the nuclear envelopes in the daughter cells following mitosis.

Intermediate filaments may contain a mixture of subunit types.

Intermediate filament proteins are capable of forming heterogenous filaments when more than one type is expressed in certain cell types. For example, keratin filaments in epithelial cells contain an equal mixture of acidic and neutral or basic keratin subunits. There are, in total, about 30 known acidic and neutral or basic keratin subtypes. Several subtypes are expressed developmentally in keratinocytes. In neurons, the light (NF-L), medium (NF-M) and heavy (NF-H) neurofilament proteins copolymerize to form neurofilaments.

Other proteins may play a role in stabilizing intermediate filaments.

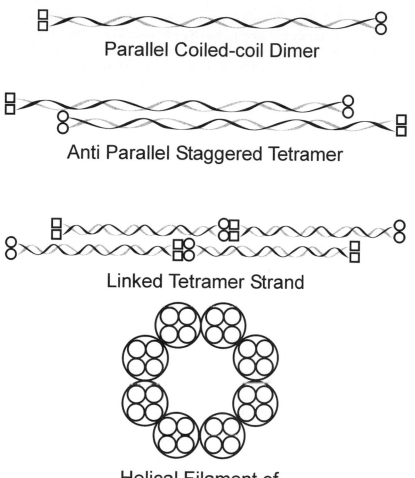

Parallel Coiled-coil Dimer

Anti Parallel Staggered Tetramer

Linked Tetramer Strand

Helical Filament of Eight Strands

Figure 3-7 Assembly and structure of intermediate filaments.
Linear IF proteins form parallel coiled-coil dimers which subsequently join to form antiparallel staggered tetramers. Tetramers join head-to-tail to form strands, eight of which form a helical IF. (Boxes denote amino termini, circles denote carboxy termini.)

Intermediate filament associated proteins (IFAPs) may stabilize filaments by cross-linking adjacent filaments, bundling filaments into parallel arrays, linking IFs to other cytoskeletal elements, or capping filament ends to prevent further assembly. Known IFAPs include *filaggrin* (associated with keratin) and *epinemin, paranemin, plectin and synemin* (associated with vimentin and/or desmin).

NUCLEUS

·

- **Eukaryotes versus Prokaryotes**
- **DNA Packaging and Packing**
- **Chromosomes**
- **Arrangement of Chromatin Within the Nucleus**
- **Fundamentals of Gene Regulation**
- **Nuclear Envelope**
- **Nucleolus and Ribosomes**
- **Transcription: One Level of Regulation of Protein Synthesis**
- **Promoter Sequences**
- **Enhancer Sequences**
- **Transcriptional Unit**
- **RNA Processing: Splicing Out the Introns and Other Modifications**
- **Levels of Intracellular Protein Regulatory Controls**
- **Nuclear Matrix**

·　·　·　·　·　·　·　·　·　·　·　·

The nucleus is the chief executive officer (CEO) of the cell. It contains the genetic material (coded in DNA) and is surrounded by a bilaminar *nuclear envelope*. It also contains the *nucleolus* located in a sea of nuclear matrix. The nuclear envelope contains pores which allow movement of molecules in a bidirectional mode. This communication with the cytoplasm is a model which would be admired by any CEO establishing an organizational plan. The nucleus regulates almost all cellular events, such as protein synthesis and cell division, although individual units are allowed some autonomy and influence nuclear decisions in an analogous manner to well run corporations.

Functionally, the nucleus is the site of transcription (transfer of genetic information to messenger RNA), extensive RNA synthesis, production of ribosomal subunits, and RNA splicing, and it serves as a communication center for information exchange with the cytoplasm.

EUKARYOTES VERSUS PROKARYOTES

In eukaryotes, the nucleus packages genetic material, coded in DNA, in a compact, organized arrangement which facilitates accurate duplication and inheritance.

The nucleus is a structure unique to eukaryotic organisms, and it distinguishes eukaryotes from prokaryotes. The sequestration of the genetic information from the cytoplasm separates RNA synthesis (a nuclear event) from protein synthesis (a cytoplasmic event). In contrast, RNA and protein synthesis occur concomitantly in the cytoplasm of bacteria and other prokaryotes.

The arrangement of the genetic material also differs between eukaryotes and prokaryotes. Prokaryotes contain a single circular molecule of DNA, whereas eukaryotes have a distinct separation of linear DNA molecules packaged into chromosomes.

DNA PACKAGING AND PACKING

The double helix of DNA is composed of a specific order of nucleotide pairs [adenosine (A) and thymidine (T), cytosine (C) and guanosine (G)], which form the letters of the genetic alphabet. Every three letters forms a *codon*, which is the code for a specific amino acid. A group of codons coding for a specific RNA molecule (*exon*) plus the associated noncoding regions (*introns*), and *noncoding regulatory segments* are known as a *gene. Multiple genes are located on a single chromosome, which consists of one entire DNA molecule and its associated packaging.*

Double helical DNA is packaged with histones and nonhistone chromosomal proteins to form chromatin fibers. (Chromatin = DNA + histones + nonhistone chromosomal proteins.)

Histones are small, positively charged, basic proteins which bind tightly to the negatively charged DNA. *Nonhistone chromosomal proteins* are negatively charged, acidic proteins (at least more acidic than the basic histones). Although the structure of nonhistone chromosomal proteins is not well understood, their functions are critical: (1) regulation of gene activity (transcription), (2) involvement in DNA replication and repair, and (3) maintenance and conversion of chromatin structure.

Histones come in two flavors, nucleosomal histones and H1 histones.

The *nucleosomal histones* are responsible for the coiling of DNA into *nucleosomes*. The nucleosomal histones are comprised of four different subtypes: H2A, H2B, H3, and H4. The H1 histones link adjacent nucleosomes.

The nucleosome is the basic functional unit of chromatin.

This basic unit consists of an *octamer of nucleosomal histones* (four pairs of specific subtypes). The next order of packing is provided by the H1 histones, which bind the nucleosomes to form the 30-nm chromatin fiber. The chromatin is elaborately folded and arranged to form another order of packing and the framework of the chromosome. The ordered packing and folding of DNA into a chromosome is shown in Fig. 4-1.

CHROMOSOMES

The chromosome is a structural unit comprising a number of not necessarily related genes and their packaging.

Each chromosome consists of a single DNA molecule plus the associated histone and nonhistone chromosomal proteins involved in packaging and gene regulation, respectively.

There are 23 pairs of chromosomes in human diploid cells. Twenty-two pairs of chromosomes are somatic chromosomes, and the additional pair consists of the sex chromosomes (XX or XY for ♀ and ♂ respectively). Haploid cells (gametes: sperm and eggs) have 22 single somatic chromosomes. There are X and Y sperm, but only X eggs.

The form of the chromosomes varies with the proliferative state of the cell.

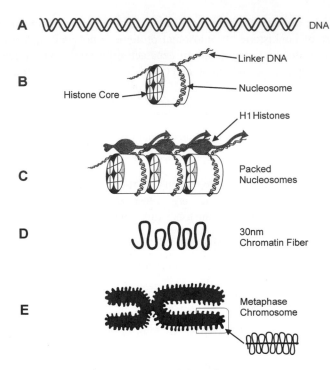

A .... DNA

B   Histone Core ——   Linker DNA
                                                     Nucleosome

C   H1 Histones
                                                     Packed
                                                     Nucleosomes

D   30nm
                                                     Chromatin Fiber

E   Metaphase
                                                     Chromosome

Figure 4-1 The packing and folding of DNA into chromosomes.
In metaphase (E) the sister chromatids are actually joined along their length. The constriction is the centromere.

In a nondividing cell the chromosome consists of a single strand. During cell division, the DNA is duplicated and two sister *chromatids* are formed. The chromosomes are most typically shown in their metaphase arrangement in which chromosomes are aligned on the cell division apparatus. At that point sister chromatids are joined along their lengths with the *centromere* consisting of the constricted region which holds the chromatids together (Fig. 4-1*E*).

Chromosomes are not just arrays of chromatin; they require specific sequences to ensure accurate replication.

The transfer of the genetic information from generation to generation requires a mechanism for making identical copies of the chromosomes for distribution to the two daughter cells which form from a parent cell. Chromosomal function requires the presence of: (1) *telomeric sequences* at the two ends of the chromosome which guarantee the integrity of the chromosome, (2) *a centromeric sequence* coding for information essential for separation of the chromosomes, and (3) *initiation replication sequences* essential for starting the copying process.

ARRANGEMENT OF CHROMATIN WITHIN THE NUCLEUS

Chromatin in the cell is found in two fundamental forms: (1) condensed and transcriptionally inactive (heterochromatin) and (2) decondensed or less condensed and transcriptionally active (euchromatin).

The biochemical structure of *euchromatin* and *heterochromatin* are different. The arrangement of the chromatin tells us about the metabolic activity of a cell. Figure 4-2 illustrates an electron micrograph of heterochromatin and euchromatin.

When chromatin becomes transcriptionally active, specific genes or regions of DNA become "unpacked" to allow binding of regulatory proteins and transcription of messenger RNA (mRNA). In the example shown in Fig. 4-2, the plasma cell is metabolically and transcriptionally active during the production of antibodies, whereas the small lymphocyte is more of a quiescent cell when passing through the bloodstream and the lymphoid organs monitoring for antigens as part of its immune surveillance function.

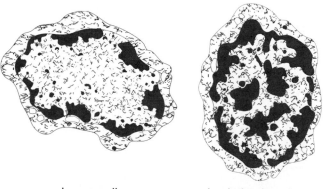

plasma cell   circulating lymphocyte

Figure 4-2 A comparison of euchromatin (plasma cell) and heterochromatin (circulating lymphocyte). Dark shading within the nucleus represents inactive heterochromatin while light shading signifies active euchromatin.

FUNDAMENTALS OF GENE REGULATION (TRANSCRIPTION)

The codon sequence in messenger RNA is complementary to the codons of DNA.

Messenger RNA is formed by the process known as transcription, and the molecules, synthesized by a specific *RNA polymerase* (polymerase II), are communication molecules representing three-letter words (codons) specific for each amino acid. The complementary nature of mRNA compared with the strand of DNA from which it was transcribed means that A replaces T, G replaces C, and C for G. *In RNA, uracil (U) is substituted for thymidine* so that U replaces A in this complementary arrangement.

Transfer RNA (tRNA) and ribosomal RNA (rRNA) are also transcribed and are considered structural components of the ribosomal machinery in contrast to the language and information carrying component provided by mRNA. The transcription process is catalyzed by the action of RNA polymerases through a similar mechanism for each of the RNAs.

Different polymerases are used for transcription of the transfer RNA (RNA polymerase III) and ribosomal RNAs (RNA polymerase I). All RNA synthesis is initiated by the binding of the specific RNA polymerase to one strand of DNA at the promoter region. This is followed by the addition of appropriate ribonucleotides to form the RNA chain with a synthetic polarity of $5' \rightarrow 3'$.

NUCLEAR ENVELOPE

The nuclear envelope is composed of inner and outer nuclear membranes and contains pores which are essential for bilateral transport of molecules to and from the cytoplasm.

The inner membrane is supported by a nuclear lamina consisting of three specific intermediate filament proteins (*lamins*). In the nondividing cell, the lamins are found in the dephosphorylated state. Phosphorylation of these proteins during the cell cycle leads to dissolution of the nuclear envelope; dephosphorylation at the completion of the cycle returns the lamins to the "resting state" where they stabilize the nuclear envelope. The outer membrane is closely associated with the rough endoplasmic reticulum of the cytoplasm.

> Nuclear pores allow passage of small molecules and ions in a nonselective manner. Larger molecules are transported by an energy-dependent mechanism.

The *nuclear pores* are formed by an octagonal arrangement of protein complexes with a central aqueous channel. *Small molecules and ions pass freely through the pore, whereas larger molecules can be selectively transported using an energy-dependent translocation process.* Messenger RNA molecules bind to a nuclear pore protein, which facilitates the passage of these molecules to the cytoplasm. Import of large molecules into the nucleus occurs in a similar fashion.

THE NUCLEOLUS AND RIBOSOMES

> The nucleolus is the center of synthesis and assembly of ribosomal subunits.

These subunits are subsequently shipped from the nucleus to the cytoplasm through the nuclear pores. All the ribosomes found in the cell are synthesized in the nucleolus except for mitochondrial ribosomes.

In humans there are 10 chromosomes which contain loops of chromatin containing *rRNA-producing DNA*. These regions or genes are known as the *nucleolar organizer regions of chromosomes*.

The drawing of an interphase nucleus (Fig. 4-3) shows the rRNA chromatin loops (nucleolar organizer region) extending into the nucleolus. In the inset, the four regions of the nucleolus are illustrated: (1) fibrillar center (tightly coiled DNA not being transcribed); (2) dense fibrillar component (sites of active transcription—RNA synthesis); (3) granular component containing maturing ribosomal units; and (4) the nucleolar matrix seen throughout the background.

As a prelude to the details of ribosomal synthesis, it is important to define the sedimentation constant. At a constant centrifugation speed, molecules or structures of similar density sediment at a constant rate expressed in Svedberg units. This S value is higher for larger molecules and represents relative molecular mass. The S values for eukaryotic ribosomal RNAs in order of increasing molecular size are 5S, 5.8S, 18S, and 28S. In eukaryotes the intact ribosomes have an 80S value composed of a 60S and 40S subunit. These are *not additive* values

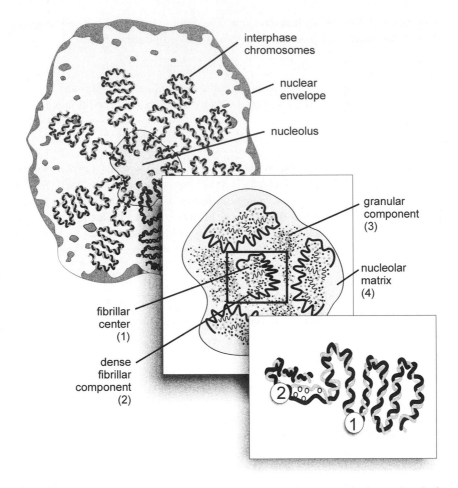

Figure 4-3 An interphase nucleus showing the RNA chromatin loops (nucleolar organizer region) extending into the nucleolus.

The middle inset is the view seen in an electron micrograph. Four regions are observed in the nucleolus: (1) fibrillar center, tightly coiled DNA not being transcribed; (2) dense fibrillar component (sites of active transcription—RNA synthesis); (3) granular component containing maturing ribosomal units; and (4) the nucleolar matrix seen throughout the background. The inset on the right illustrates a high magnification view of the nucleolar organizer region and corresponds with the dense fibrillar component, which is an area of active transcription (2) and the fibrillar center which is an area which is transcriptionally inactive (1).

because they represent independent sedimentation constants. The table below lists the S values for the eukaryotic and prokaryotic ribosomes and their subunits.

CLASSIFICATION OF RIBOSOMES AND THEIR SUBUNITS BASED ON THEIR SEDIMENTATION CONSTANTS

TYPE	RIBOSOME	LARGE SUBUNIT	rRNAS OF LARGE SUBUNIT	SMALL SUBUNIT	rRNAS OF SMALL SUBUNIT
Eukaryote	80S	60S	5S, 5.8S, 28S	40S	18S
Prokaryote	70S	50S	23S, 5S	30S	16S

Transcription and processing of pre-RNAs occurs in the nucleolus.

The first step in eukaryotic ribosomal synthesis is transcription by RNA polymerase I to form the so-called pre-RNA (45S) transcript. The 5.8S, 18S, and 28S sequences are processed from that original pre-RNA transcript.

Proteins associated with the rRNAs are synthesized in the cytoplasm and transported to the nucleolus for assembly of ribonucleoprotein. The 5S rRNA is synthesized outside the nucleolus, but transported into it for assembly of the large subunit.

A specific RNA polymerase (RNA polymerase III) catalyzes the transcription of the pre-5S RNA, which is processed and subsequently transported to the nucleolus where it is combined with the 5.8S and 28S to form the large subunit.

Proteins synthesized in the cytoplasm are added to these rRNAs to form ribonucleoprotein complexes. Both the large and small subunits contain proteins in addition to the rRNAs. These patterns are similar in prokaryotes although their S values differ.

Functional ribosomes exist only after transport to the cytoplasm and activation.

This process separates functional ribosomes from the incompletely processed RNA found in the nucleus.

> Regions of the nucleolus contain distinct macromolecules involved in the process of ribosomal production and maturation. The shape and size of the nucleolus varies with the activity of the cell.

Four regions have been identified within the nucleolus as described earlier and shown in Fig. 4-3: (1) a fibrillar center which is pale staining, (2) a dense fibrillar component, (3) a granular component, and (4) the nucleolar matrix. Activity of the cell and size of the nucleolus are directly related. The granular component is the region which shows the greatest increase in size during increased cellular activity reflecting increased ribosomal gene transcription as well as increased ribosomal synthesis. Large neurons are usually very active transcriptionally and are noted for their large, prominent nucleoli and euchromatic nuclei.

TRANSCRIPTION: ONE LEVEL OF REGULATION OF PROTEIN SYNTHESIS

Transcription of messenger RNA from DNA occurs within the nucleus and is the first step in converting the information of the genome, retained in DNA, to protein.

> Transcription consists of two primary steps: (1) activation defined as the biochemical and morphologic changes in chromatin structure associated with transcription and (2) modulation, the turning on of transcription by gene regulatory proteins known as transcription factors, which bind to DNA, initiating the process.

Activation involves the conversion of chromatin from one state to the other. Morphologically this is seen as the conversion from heterochromatin to euchromatin (discussed above in the section on the arrangement of chromatin in the nucleus). There are also biochemical modifications which are poorly characterized.

Gene regulatory proteins are classified into a number of categories based on their structural arrangement (motif). These proteins must recognize distinct sequences of double-stranded DNA and in some cases mediate their own dimerization.

The classifications of gene regulatory proteins include (1) helix-turn-helix (two α helices connected by a turn composed of a short amino acid chain), (2) homeodomain (helix-turn-helix arrangement with a specific amino acid sequence in the turn), (3) zinc finger motifs (zinc forms a structural component and uses an α helix to recognize the major groove in DNA), (4) the leucine zipper motifs (two α helices held together by the hydrophobic side chains on leucines), and (5) helix-loop-helix motifs (two α helices connected by a protein loop which is structurally flexible). The leucine zipper and helix-loop-helix proteins mediate both DNA binding and dimerization. The process of dimerization appears to enhance the strength of protein-DNA binding and most often occurs at a site on the regulatory protein separate from the DNA binding site.

Gene regulatory proteins (transcription factors) recognize two classifications of DNA sequences: (1) promoter and (2) enhancer sequences.

PROMOTER SEQUENCES

The promoter region is the site of initiation of RNA polymerase activity on the DNA.

There are two major DNA "core" promoter sequence sites: (1) the TATA sequence, an area of high thymidine (T) and adenine (A) bases, therefore the name TATA, and (2) the CAAT sequence, consisting of a high concentration of cytosine (C) as well as adenine and thymidine; resulting in the term CAAT.

The promoter–regulatory protein binding interaction may function through two proposed mechanisms: (1) modification of the DNA to facilitate the binding of the appropriate RNA polymerase or (2) direct protein-protein binding of the gene regulatory protein to the RNA polymerase.

ENHANCER SEQUENCES

The enhancer sequences influence the rate of transcription.

Enhancer sequences are recognized and bound by gene regulatory proteins and may be located upstream (toward the 5' end of the DNA being read) or downstream (toward the 3' end). The enhancers function independently of polarity and may be located thousands of base pairs from the promoter.

TRANSCRIPTIONAL UNIT

A transcriptional unit has been defined from the start site (promoter on the DNA) to the termination site, where the newly formed RNA (primary) transcript is released from the RNA polymerase. The growing RNA transcripts remain attached to the RNA polymerase molecules as they travel along the DNA. This results in the forming of branches (RNA polymerases plus attached RNA transcripts) and the trunk (DNA double helix) of a pine tree-like pattern similar to that of a Christmas tree.

Transcription is only one mechanism for cellular control over protein production and synthesis. It is the first level in the process of transfer of information from the genome to protein. During transcription, the genome is read; however, subsequent steps allow for variation and modification of the protein.

RNA PROCESSING: SPLICING OUT THE INTRONS AND OTHER MODIFICATIONS

The mRNA undergoes a process in which the noncoding sequences (*introns*) are removed from the message and only the coding sequences (*exons*) remain. The removal of the introns and joining of the coding regions together are called *RNA splicing*. This system of reading the genome results in differences between the primary transcript read from the DNA and the mRNA which is used in the subsequent step of translation in the cytoplasm by ribosomes.

RNA splicing is carried out by small nuclear ribonucleoproteins (snRNPs).

The cute term snRNPs, pronounced *snurps* (not related to the Smurfs of cartoon fame), has been used for this acronym. The snRNPs are ribonucleoprotein complexes and are analogous to ribosomes in structure (combination of RNA and protein), but about 1/20 the molecular weight.

The snRNPs recognize specific RNA sequences and are involved in the formation of an intermediate of splicing known as the *spliceosome*. The introns are excised from the spliceosome in the form of a lariat and undergo subsequent breakdown in the nucleus. Other modifications of the primary transcript include two types of additions, one at each end of the mRNA: (1) a 5' cap instrumental later in the initiation of protein synthesis during translation in the cytoplasm and (2) the addition of a polyadenine (poly A) tail at the 3' end. The poly A tail serves a number of important functions. It enhances the stability of messenger RNA, facilitates nuclear export, and recognizes the ribosomes. These modifications precede RNA splicing.

LEVELS OF INTRACELLULAR PROTEIN REGULATORY CONTROLS

RNA splicing represents one possible site of modification of the protein program. Alternate splicing of RNA can result in the production of different protein transcripts. This can lead to development of severe genetic diseases. In the case of some β thalassemias, a disease in which patients suffer from abnormally low hemoglobin levels, there is a mutation (heritable change in the nucleotide sequence of the DNA) or mutations in the introns of the β-globin gene. This results in impaired splicing of the mRNA. Interestingly, there are also other β thalassemias in which a single base substitution in the codon region creates a stop codon or frame shift leading to a nonsense mutation in the coding region of the β-globin gene. Therefore, both introns and exons are important in the cascade of nuclear RNA processing events.

Other factors such as the rapidity of transport from the nucleus, stability of mRNAs, as well as regulation of translation and posttranslational activation and modification represent different levels of protein synthetic controls within the cell. The second major step in the transfer of information is the process of translation in which the mRNA information is transferred to protein at the ribosome (Chap. 5, "Protein Synthesis").

NUCLEAR MATRIX

In a similar fashion to the cytosolic matrix, the nucleus also has a matrix which has a distinct organization.

The nuclear matrix contains proteins which bind DNA. These binding proteins are important to the positioning of chromosomes.

The changes which occur in the nucleus during cell division and the intricacies of the replication process which allows accurate duplication of the genetic material are discussed in Chap. 10.

PROTEIN SYNTHESIS

·

- **Ribosomal Components**
- **Translation of Messenger RNA**
- **Signal Hypothesis**
- **Fate of Newly Synthesized Proteins**

·   ·   ·   ·   ·   ·   ·   ·   ·   ·   ·   ·

Cells literally synthesize everything from soup (membrane lipids) to nuts (ribosomes). However, the majority of the cellular synthetic machinery is dedicated to the production of peptides and proteins. Among the wide variety of proteins synthesized by most cells are cytoskeletal proteins, transmembrane receptors and channels, regulatory proteins, metabolic enzymes, extracellular matrix proteins, and hormones. Proteins are synthesized on ribosomes which *translate* the coded sequence of nucleotides that make up messenger ribonucleic acid (mRNA) which was *transcribed* from the deoxyribonucleic acid template of the genes. Within the amino acid sequence coded by mRNA are one or more chemical signals which determine the destination and configuration of each protein. Following translation, proteins may be further modified, for example, by the addition of carbohydrate residues or lipid anchors.

RIBOSOMAL COMPONENTS

The RNA components (rRNA) of ribosomes are transcribed in the nucleolus and transported through nuclear pores to the cytoplasm.

The synthesis and composition of ribosomal components have previously been described in Chap. 4, "Nucleus." Basically, ribosomes consist of a large (60 S) subunit (28, 5.8, and 5 S rRNAs and numerous proteins) and a smaller (40 S) subunit (18 S rRNA and numerous proteins) (Fig. 5-1).

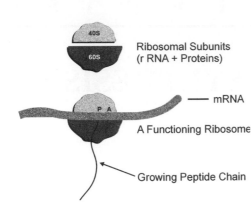

Ribosomal Subunits
(r RNA + Proteins)

mRNA

A Functioning Ribosome

Growing Peptide Chain

Figure 5-1 Formation of ribo-somes from subunits in the pres-ence of mRNA.
Small (40S) and large (60S) ribo-somal subunits combine in the pres-ence of mRNA to begin protein syn-thesis. P and A are binding sites for aminoacyl-tRNA.

The location of ribosomes within the cytoplasm reflects the fate of newly synthesized proteins.

Proteins destined for use in the cytoplasm (soluble proteins and peripheral membrane proteins), as well as proteins targeted to mitochondria or the nucleus, are synthesized on ribosomes floating "free" within the cytoplasm. Proteins des-tined for export (secretion) (Chap. 6), distribution to specific organelles such as lysosomes (Chap. 6), or insertion into various membrane compartments (Chap. 1), are synthesized on ribosomes associated with the cytosolic surface of the endo-plasmic reticulum or rough ER (RER) (Chap. 2). As shown later, it is actually the protein being synthesized which determines the type of ribosome.

TRANSLATION OF MESSENGER RNA

Messenger RNA (mRNA) contains codons of three nucleotides each, which determine the sequence of amino acids making up a protein.

The sequential series of amino acids to be combined to form each specific protein is determined by *codons* in the mRNA. Each codon consists of three nucleotides grouped in a specific order and specifying a particular amino acid. Since four different nucleotides [uracil (U), adenine (A), cytosine (C) and gua-nine (G)] are found in mRNA, there are 64 (4×4×4) possible combinations of nucleotides in each three-nucleotide codon. Obviously, since there are only 20 amino acids, some amino acids are coded for by more than one codon. Generally, the first two nucleotides of a codon are stable, whereas the third may vary.

Transfer RNAs (tRNAs) carry specific amino acids to the ribosomal-mRNA complex for addition to the growing amino acid (polypeptide) chain.

Transfer RNAs are transcribed in the nucleus and exported to the cytoplasm. Within their nucleotide sequence is a group of three nucleotides (*anticodon*) which is complementary to a codon in the mRNA being translated. The tRNAs also carry the amino acids (aminoacyl-tRNA) specified by each codon of the mRNA (Fig. 5-2). Hydrogen bonding between the complementary nucleotide pairs of the codon-anticodon position the aminoacyl-tRNA so that the amino acid can be added to the growing peptide chain.

The process of ribosomal protein synthesis consists of three steps: initiation, elongation, and termination.

Initiation begins with the binding of the small ribosomal subunit to the 5' end of the mRNA. The small ribosomal subunit then scans the mRNA (channel surfing) toward the 3' end until it finds the *initiation codon (AUG)*. At this point, the *initiator tRNA* (with an anticodon to AUG) binds to the ribosomal-mRNA complex (Fig. 5-2A). The initiation codon always codes for methionine-tRNA, and thus the first amino acid in a newly synthesized protein is always methionine. The special initiator tRNA is necessary to fool the ribosome into believing it already has a growing protein at one of its binding sites (see below). Following binding of the initiator tRNA, the large ribosomal subunit is added to the complex. *Special initiation factors control the rate of initiation.*

Additional amino acids are added to the growing protein during the *elongation* phase.

There are actually two binding sites for tRNA on each ribosome. The A binding site binds each new aminoacyl-tRNA *after* the initiator tRNA binds to the P site. Ribosomal enzymatic activity (peptidyl transferase) then catalyses the formation of a peptide bond between the amino acid at the P site and the new amino acid at the A site. The empty tRNA at the P site is then released, and the ribosome shifts exactly three nucleotides (one codon) down the mRNA (toward the 3' end) (Fig. 5-2 B–D). This accomplishes two things: (1) it shifts the polypeptide chain at the A site to the P site and (2) it exposes the next codon at the A site. The next coded aminoacyl-tRNA binds at the A site (Fig. 5-2E), and the process is repeated. This explains why a special initiator tRNA is required to bind first at the P site (instead of at the A site) to mimic a tRNA attached to a growing polypeptide chain.

A special nucleotide sequence in the mRNA signals the end of translation.

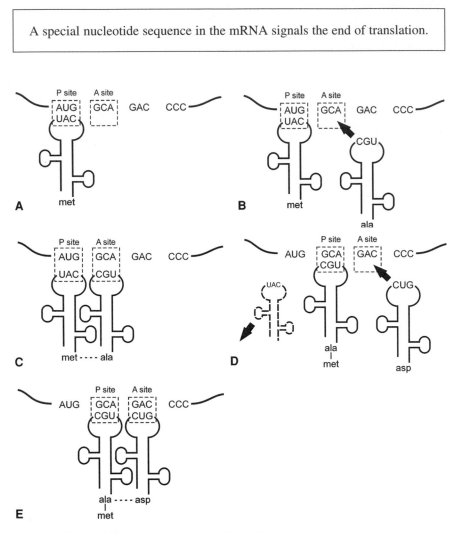

Figure 5-2 Synthesis of a four amino acid peptide.
A. The first aminoacyl-tRNA binds to the P site. The first amino acid is always methionine. *B, C.* The second aminoacyl-tRNA binds at the A site. A peptide bond forms between the two amino acids. *D.* The empty tRNA is released, and the ribosome shifts one codon toward the 3′ end of the mRNA, delivering the tRNA with the growing peptide chain attached to the P site and exposing the next codon at the A site. *E.* The next aminoacyl-tRNA attaches at the A site and the process in C and D is repeated.

The *termination* of protein synthesis occurs when a "stop" codon (either UAA, UAG, or UGA) reaches the A site. This signals the enzymatic release of the completed polypeptide chain, release of the empty tRNA at the P site, and dissociation of the ribosomal subunits from the mRNA (Fig. 5-2 *H, I*).

Many ribosomes may read the same mRNA simultaneously.

As ribosomes begin to translate mRNA at the 5' end and read toward the 3' end, new ribosomes may begin at the 5' end as soon as space is available. These aggregations of ribosomes on a single strand of mRNA are termed *polysomes* and may be either free (Fig. 5-3) or membrane bound. The "older" ribosomes nearer the 3' end always demonstrate longer polypeptide chains than the "newcomers" near the 5' end.

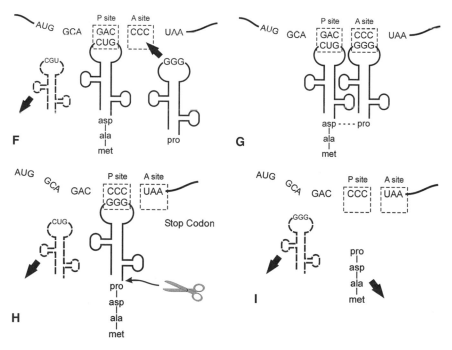

Figure 5-2 (*continued*)
F, G. The process is repeated for the final amino acid. *H, I.* The ribosome shifts another codon and recognizes a "stop" codon at the A site, resulting in enzymatic cleavage of the peptide chain from the final tRNA and detachment of both the peptide and tRNA from the ribosomal-mRNA complex.

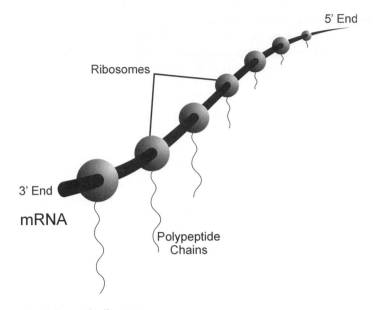

Figure 5-3 A free polyribosome.
Many ribosomes are reading the same mRNA.

Following synthesis and release from the ribosomes, proteins may be modified and folded.

One or more of the amino acids in a newly synthesized protein may be transformed into a different amino acid by any number of chemical modifications including methylation, sulfation, carboxylation, and hydroxylation. One example would be the hydroxylation of proline residues in newly synthesized collagen. Other posttranslational modifications include the addition of carbohydrate residues or the addition of lipid moieties which anchor the protein to a membrane. Proteins may also be enzymatically cleaved at either end or even in the middle. The processing of atrial natriuretic peptide, a peptide hormone with antihypertensive and natriuretic-diuretic properties, is a good example of end cleavage. In the cardiac atria, ANP is synthesized as a 152 amino acid polypeptide, including a 26 amino acid signal sequence (prepro-ANP). The signal sequence is cleaved immediately, and the peptide is stored in its proform (pro-ANP). On secretion, pro-ANP is further cleaved to a 28 amino acid circulating form. On the other hand, preopiomelanocortin (POMC) is cleaved internally to a variety of peptides including adrenocorticotrophic hormone (ACTH) and melanocyte stimulating hormone (MSH).

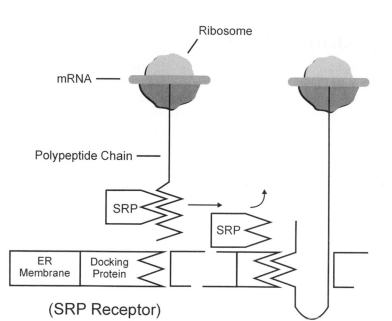

Figure 5-4 Role of the signal recognition particle (SRP).
Peptides to be synthesized on membrane-bound ribosomes (RER) possess a signal sequence at their amino terminus which is recognized by the SRP and binds the ribosome and polypeptide chain to the RER membrane via a docking protein (SRP receptor). The hydrophobic signal sequence on the polypeptide is inserted into the lipid bilayer of the membrane, and the polypeptide chain elongates into the RER lumen. Subsequently, enzymatic cleavage of the signal sequence releases the finished polypeptide.

Chemical modification affects the final folding of proteins and vice versa. In addition, other proteins may assist when specific folding arrangements are required.

Proteins like to assume the lowest possible energy state compatible with function (much the same as when, after a hard day, someone goes home and curls up into the fetal position on a sofa). The chemical modifications which proteins undergo during posttranslational processing enable and stabilize the proper folding conformation. Molecular *chaperones* help newly synthesized proteins maintain folded states which may be at intermediate energy levels necessary for insertion of proteins through membranes. Among the chaperones is a class of proteins, the *chaperonins*, which include the heat shock proteins such as hsp 70.

SIGNAL HYPOTHESIS

As mentioned previously, proteins destined for use in the cytoplasm or for import into the nucleus or mitochondria are synthesized on free ribosomes, whereas those destined for export, distribution to lysosomes, or incorporation into membranes are synthesized on ribosomes associated with the endoplasmic reticulum (RER). How does a protein "know" on which type of ribosome it is to be synthesized?

All proteins synthesized on ribosomes associated with the RER contain a signal sequence of amino acids.

Actually, the synthesis of all proteins begins on "free" ribosomes. However, forming proteins which possess a special *signal sequence* direct their ribosomes to attach to the membrane of the endoplasmic reticulum. The signal sequence, usually found at the N terminal of a forming protein, is not dependent on a specific sequence of amino acids but rather on regions with high concentrations of hydrophobic residues. The N-terminal portion of the protein is the first part synthesized and is believed to extend through a channel in the large ribosomal subunit.

A soluble signal recognition particle binds to the N-terminal signal sequence and to the ribosome.

The signal recognition particle (SRP) consists of six proteins and a small (7 S) segment of RNA (scRNA). The binding of SRP to the nascent protein and ribosomal complex temporarily inhibits further growth of the protein. Once bound to the signal portion of the growing protein and to the ribosome, the SRP also binds to a receptor (docking protein) on the cytoplasmic face of the ER membrane (Fig. 5-4). The SRP is subsequently released, and the hydrophobic signal sequence is inserted into the lipid bilayer of the ER membrane. Elongation of the peptide resumes, and the growing amino acid chain is shuttled to the lumen of the ER through as yet poorly defined protein translocators in the ER membrane. For soluble, non-transmembrane proteins, a peptidase cleaves the mature peptide from the signal peptide sequence and releases it to the ER lumen.

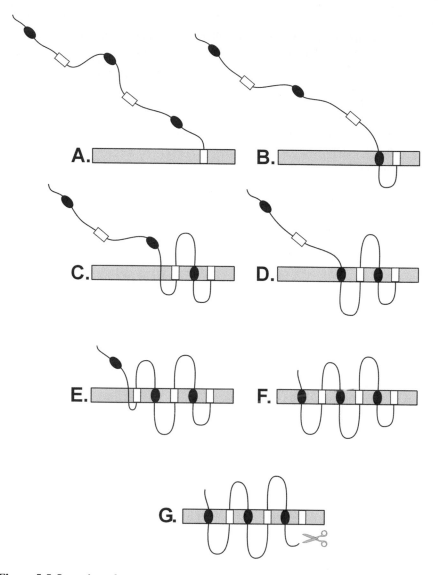

Figure 5-5 Insertion of a transmembrane protein into the membrane of the RER.
A. The initial "start" (signal) sequence (white rectangle) is inserted in the RER membrane.
B. As described in Figure 5-4, the polypeptide chain is elongated into the RER lumen until
a "stop" sequence (black oval) is reached. C. The next start sequence is inserted into the
membrane. D–F. The sequence is repeated until the final stop sequence is reached and the
polypeptide is enzymatically cleaved from the initial start sequence (G).

Transmembrane proteins contain alternating "start" and "stop" signal sequences which determine the number of transmembrane segments.

Both start and stop signal sequences consist of 20 to 30 hydrophobic amino acids in an α-helical conformation. Once a stop sequence is reached and the newly synthesized protein is inserted through the protein translocator, its hydrophobic nature directs its incorporation into the lipid bilayer of the membrane. Rather than being released, the SRP scans the amino acid sequence for the next start signal to insert in the membrane. This leaves a loop of predominantly hydrophilic residues on one side of the RER membrane. The start sequence then directs translocation of the next hydrophilic segment across the membrane until the next stop sequence is reached (Fig. 5-5).

FATE OF NEWLY SYNTHESIZED PROTEINS

The authors have attempted not to burden this magical mystery tour through the cell with any raw experimental data or long, boring narratives of experiments which only a few students (unfortunately) have time to read. However, the experiments of Jamieson and Palade (1971) were so classic and basic to the understanding of protein synthesis, processing, and storage-export (exocytosis) that readers should know their fundamental design and results.

Jamieson and Palade incubated slices of pancreas in a medium containing a radioactive amino acid which would be taken up by the cells and incorporated into newly synthesized proteins. Pancreatic acinar cells synthesize and secrete numerous digestive enzymes (proteins) in a regulated pathway (see Chap. 6). They represent the quintessential polarized exocrine cell. After a brief exposure to the radioactive amino acid (pulse), the slices were washed and the incubation medium was replaced with one containing an excess of unlabeled amino acids (chase). This classic pulse-chase experiment (Fig. 5-6) enabled the investigators to follow and locate the labeled amino acids as they proceeded through the various processes of protein synthesis. Using the technique of autoradiography in which very thin sections of tissue are cut, coated with a photographic emulsion sensitive to the radioactive label, exposed, and viewed in an electron microscope, Jamieson and Palade were able to detect the location of labeled amino acids incorporated into new proteins at various time periods after the initial pulse.

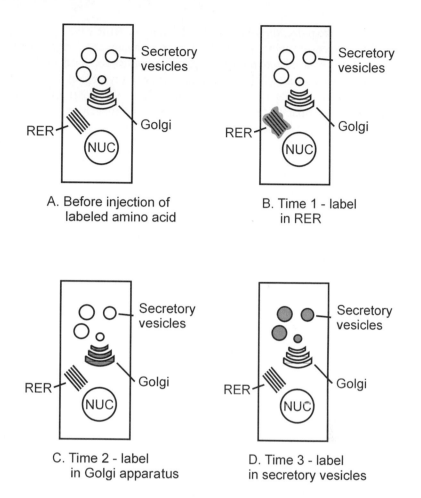

Figure 5-6 The protein synthesis and secretion pathway in a typical exocrine cell.
A. Prior to injection of radiolabeled amino acid, cells show no labeling of RER, Golgi apparatus, or secretory vesicles. *B.* Within a few minutes of injection of labeled amino acid followed by a "chase" of a high concentration of unlabeled amino acid, labeling is found only over the RER. *C.* After a few more minutes, labeling has vacated the RER and has moved on to the Golgi apparatus. *D.* At a subsequent time, labeling has left the Golgi apparatus and is now found in secretory vesicles moving toward the apical plasma membrane. This experiment clearly demonstrates that proteins for export are synthesized on the RER and shuttled to the Golgi apparatus where they are packaged into secretory vesicles for subsequent secretion (exocytosis).

Within 3 min of the pulse, the label was localized in the RER where it was incorporated into nascent proteins as described in this chapter. At longer durations after the pulse, the label was present in proteins being processed in the Golgi apparatus and subsequently being stored in secretory (zymogen) vesicles. The routing of newly synthesized proteins from the RER to the Golgi apparatus and then to a number of final destinations is discussed in Chap. 6.

INTRACELLULAR TRAFFICKING

•

- **Endocytosis**

- **Exocytosis**

- **Transport to Lysosomes**

- **Transport to the Nucleus**

- **Transport to Mitochondria**

• • • • • • • • • • • •

All cells are, to one extent or another, engaged in the import-export business and thus require shipping and receiving departments as well as an efficient system of interdepartmental communication (cellular E mail). The overriding concept of this intracellular business is the trafficking of materials from compartment to compartment within the cell, as well as to and from the external environment, *via membrane-bound vesicles. Endocytosis* is the import side of the operation, and *exocytosis* is the export side. The balance between export and import is exemplified by the conservation and reutilization of membrane components. This results in a highly efficient system with little waste of energy.

ENDOCYTOSIS

Phagocytosis is the mechanism by which specialized cells import foreign objects such as bacteria for subsequent digestion into usable fragments.

Once bacteria and other foreign particles enter the body, they are generally coated by *antibodies* secreted by plasma cells derived from lymphocytes. Blood cells (neutrophils) and monocyte-derived macrophages in the tissues possess cell surface receptors which recognize the *Fc subunit* of these antibodies. Binding of the Fc subunits to the receptors initiates *evagination* of the cell membrane, which eventually surrounds and engulfs the foreign particle as a large, membrane-bound vesicle. The *phagocytic vesicle* then merges with a lysosome containing multiple hydrolyzing enzymes, and the degradation of the foreign object ensues.

Pinocytosis is the mechanism by which small objects (e.g., fluids, soluble proteins, and receptor-ligand complexes) are imported into the cells of multicellular organisms.

Pinocytosis ("cell drinking") involves the *invagination* of plasma membrane to produce small, membrane-limited vesicles. There are two forms of pinocytosis: *fluid-phase endocytosis* and *receptor-mediated endocytosis*. Fluid-phase endocytosis is *nonselective* and results in the formation of vesicles containing extracellular fluid and all its soluble contents.

Receptor-mediated endocytosis is selective. Specific ligands binding to receptors associated with "coated pits" of the plasma membrane are internalized in this manner.

The association of a high density of specific receptors with *coated pits* of the plasma membrane allows the *concentration of ligand* within the *endocytotic vesicles* to be as high as 1000-fold greater than in the extracellular fluid. The mechanism by which receptors are concentrated in the coated pits is unknown; some receptors are prepositioned in the pits before binding ligand, whereas others must bind ligand before entering the pits.

Membrane pits involved in selective transport are lined with clathrin.

Clathrin forms the "coat" of coated pits involved in selective endocytosis as well as the coat of secretory vesicles of the regulated secretory pathway of exocytosis (to be discussed later). Clathrin subunits consist of three heavy and three light chains arranged as a bent tripod, or *triskelion* (Fig. 6-1). Triskelions aggregate on the *cytoplasmic surface* of a membrane pit or a nascent endocytotic vesicle in a pattern of interlocking pentagons and hexagons. The pattern resembles Buckminster Fuller's geodesic dome with the scaffolding on the outside. *The process of clathrin assembly drives the formation of vesicles and does not require additional energy.*

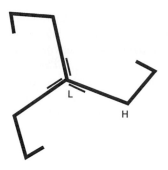

Figure 6-1 The clathrin triskelion. Triskelions consist of three heavy (H) and three light (L) clathrin subunits. Triskelions assemble on the cytoplasmic surface of membranes in a geometric pattern reminiscent of a soccer ball and are anchored to transmembrane proteins by adaptins.

Clathrin anchors the transmembrane receptors at the site of the forming endocytotic vesicle.

The cytoplasmic ends of transmembrane receptor-ligand complexes are connected to the clathrin coat by a variety of intermediary proteins called *adaptins*. The activation and binding of adaptins to the receptor-ligand complexes initiates or enhances formation of the clathrin coat.

The clathrin coat is disassembled immediately after release of the new vesicle into the cytoplasm.

Clathrin is apparently required only for formation and budding of vesicles from cell membranes and not for subsequent movement or fusion with other membranes. The "decoating" requires ATP and is carried out by an ATPase of the *heat shock protein family (hsp 70)*. The newly formed vesicles are then transported along microtubules and fuse with *early endosomes*, another membrane-bound compartment.

Endosomes (and lysosomes) contain proton pumps in their membranes which transport H^+ into the lumen and acidify the contents.

The process of acidification begins in early endosomes and results in the dissociation of some ligands from their receptors. The processing of cholesterol (low-density lipoprotein, LDL) bound to its surface receptor (LDL-LDL receptor complex) is a well known example. The dissociated LDL is shuttled to late endosomes where the import of *acid hydrolases* from the *trans*-Golgi network converts these compartments to lysosomes where the LDL is degraded. *The early endosome membrane with unoccupied LDL receptors is recycled back to the plasma membrane.* Other receptor-ligand complexes entering early endosomes are not sensitive to pH, and the entire complex is shuttled to the lysosomal compartment and only the endosomal membrane is recycled.

EXOCYTOSIS

Protein and membrane synthesis have been covered in previous chapters. Proteins for export (exocytosis) leave the RER in membrane-bound *transport vesicles* which fuse with the *cis* face of the Golgi apparatus (*cis-Golgi network*, CGN). Subsequent processing of the export protein through the lamellae of the Golgi apparatus also utilizes vesicular transport. At the *trans* face of the Golgi apparatus (*trans-Golgi network*, TGN), proteins are sorted according to destination and ferried by vesicles to the plasma membrane or lysosomes (Fig. 6-2).

Nonselective transport pathways utilize coatomers (COPs) rather than clathrin as vesicle-coating proteins.

Nonselective transport (or "default") pathways require no signal element and include vesicular shuttling (1) between the RER and CGN, (2) within the Golgi apparatus, and (3) via the constitutive (unregulated) secretory pathway from the TGN to the plasma membrane. *Coatomer* complexes of five to eight subunits bind to the vesicular membranes via a monomeric G protein (ARF, adenosine diphosphate ribosylation factor) which is inserted into the vesicular membrane when activated by binding GTP. Unlike clathrin-coated vesicles, *the coatomer coat does not dissociate immediately after vesicular formation,* but remains until the vesicle docks with the target membrane. A GTPase-activating protein in the target membrane initiates the hydrolysis of GTP. (Review the structure and function of G proteins in Chap. 9). This results in the dissociation of ARF and coatomer from the vesicle membrane.

Proteins must be properly folded to be exported.

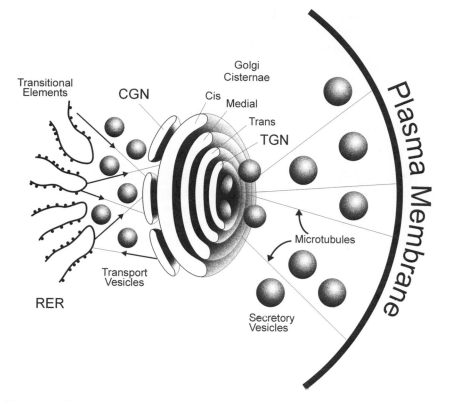

Figure 6-2 Pathways of exocytosis.
Proteins for export are synthesized in the RER and collected in transport vesicles for transport along microtubules to the *cis* face (CGN) of the Golgi apparatus. Transport between the stacks of Golgi cisternae also uses transport vesicles. Secretory vesicles leaving the *trans*-Golgi network (TGN) may enter the constitutive or regulated exocytotic pathways.

Quality control of exported proteins begins in the RER. Proteins with defects such as amino acid deletions or improper sequences which result in incorrect folding are excluded from the transport vesicles shuttling to the CGN and are degraded within the RER. In addition, any ER-specific proteins (containing an ER retention sequence) which escape the ER "prison" and reach the CGN, are "arrested" by binding to receptors in the CGN membrane and are returned to the ER via transport vesicles (also see Fig. 2-2, KDEL). Even at the subcellular level, you can run but you can't hide.

Proteins are sorted in the *trans*-Golgi network.

After reaching the CGN, membrane-bound and soluble proteins are shuttled sequentially through the Golgi cisternae by coatomer-coated, nonspecific transport vesicles. When reaching the TGN, proteins are sorted according to destination and function.

Proteins lacking specific signal sequences are ferried in coatomer-coated vesicles directly to the cell membrane.

Proteins not specifically targeted to secretory vesicles (the regulated pathway) or diverted to lysosomes enter the "default," or constitutive, secretory pathway as a continuous stream of small, coatomer-coated vesicles leaving the TGN (Fig. 6-3). When docking with the plasma membrane, they contribute new phospholipids and cholesterol to the plasma membrane while releasing their soluble protein contents to the exterior. *Components of the extracellular matrix are secreted in this fashion.*

Proteins containing specific signal sequences enter the regulated secretory pathway.

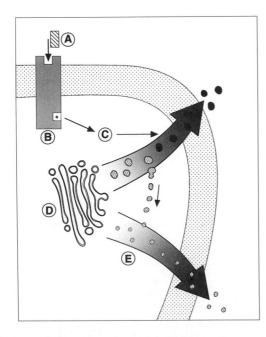

Figure 6-3 Regulated and constitutive secretory pathways. The regulated secretory pathway (A-B-C) demonstrates the secretion of a stored product following binding of a ligand (A) to its receptor (B). The product was stored in clathrin-coated vesicles budding from the Golgi apparatus (D). The constitutive or default pathway (E) represents unstimulated release of product and is the major pathway for shuttling lipids and integral membrane proteins to the cell membrane.

As yet unrevealed forces or signals cause aggregation of these proteins in the vicinity of forming clathrin-coated secretory vesicles in the TGN. After budding from the TGN, secretory vesicles lose their clathrin coating and travel along microtubules to the vicinity of the cell membrane using motor proteins of the kinesin family.

> Regulated secretion is triggered by external agents.

Chemical agents such as hormones and neurotransmitters bind to trans-membrane receptors on the plasma membrane and initiate a series of events usually leading to a localized and transient elevation of Ca^{2+} concentration. This leads to penetration of the cytoskeleton and docking of the secretory vesicle with the plasma membrane followed by membrane fusion and release of the vesicular contents to the exterior.

> Docking of vesicles at target membranes and subsequent membrane fusion follow a common pathway regardless of the origin of the vesicle or the location of the target.

The mechanisms regulating docking and fusion are similar whether the vesicle arises from the ER and docks at the CGN, arises from one Golgi cisterna and docks at another, or buds from the TGN and docks with the plasma membrane. The process of vesicle docking and fusion has been extensively studied in neurons where synaptic vesicles dock and fuse with the membrane of the presynaptic terminal.

> The first requirement for docking is removal of the vesicular coat.

As may be recalled from earlier in this chapter, removal of clathrin from endocytotic or secretory vesicles occurs immediately after budding from the parent membrane. The removal of coatomer from vesicles in nonselective pathways occurs later.

> Secretory vesicles penetrate the cytoskeleton and become linked to the plasma membrane through interactions with spectrin or related proteins.

As noted in Chap. 3, "Cytoskeleton," members of the spectrin protein family form an integral part of the cytoskeletal latticework supporting the cell membrane. As the vesicles approach the spectrin cytoskeleton, *synapsins* on their membranes bind to spectrin tails, displacing actin.

A delicate ballet of binding and recognition steps completes docking of the vesicle and initiates fusion of the membranes.

This is analogous to the secret fraternity handshake. Vesicular and target membranes possess specific receptor proteins, [SNARE-v (VAMP) and SNARE-t, respectively]. VAMP and *synaptotagmin* (calcium-sensitive) on the vesicular membrane recognize and bind two forms of SNARE-t, *syntaxin and SNAP-25* on the target membrane, completing docking of the vesicle. The accuracy of the docking between SNAREs is verified by monomeric GTPases which hydrolyze Rab-GTP preceding membrane fusion (Fig. 6-4).

A receptor protein, SNAP (soluble NSF attachment factor), binds to the SNARE handshake.

Once this is accomplished, NSF (*N*-ethylmaleimide sensitive factor) binds to SNAP. NSF is an ATPase. Hydrolysis of ATP results in release of both NSF and SNAP from the SNARE complex and initiation of Ca^{2+}-dependent membrane fusion.

Membrane fusion requires a fusion protein (fusigen).

Researchers have hypothesized that the fusigen overcomes the hydrophobic forces on the membrane surface and allows the hydrophobic portions of the outer and inner membrane leaflet of both the vesicle and the target membrane to merge.

Following exocytosis, membrane components are recycled to the TGN.

The continuous addition of membrane from both the constitutive and regulated secretory pathways would rapidly increase the surface area of the plasma membrane if equal amounts of membrane were not removed from the plasma

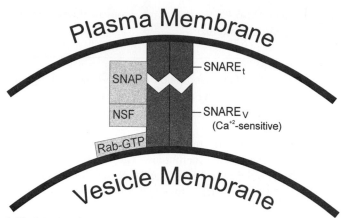

Figure 6-4 Vesicle docking.
SNARE receptors on the vesicular (v) and target (t) membranes initiate docking via a Ca^{2+}-dependent mechanism. Docking is secured by Rab-GTP. Fusion of the two membranes is initiated through the action of NSF/SNAP and a fusigen.

membrane by endocytosis. In this way, protein and phospholipid membrane components are returned to the TGN for reuse in the exocytotic pathways.

TRANSPORT TO LYSOSOMES

Lysosomes are involved in the degradation of products delivered to them by (1) phagocytosis, (2) fluid- and receptor-mediated endocytosis, (3) autophagic vesicles containing worn-out organelles and other cellular debris, and (4) a direct pathway involving specific signals on proteins free in the cytoplasm. Lysosomal enzymes, as well as membrane proteins mediating lysosomal function (e.g. transporters and proton pumps), are targeted to late endosomes via clathrin-coated transport vesicles originating from the TGN.

The targeting signal for lysosomal enzymes is mannose-6-phosphate.

Within the CGN, specific mannose residues of glycoproteins destined to become lysosomal acid hydrolases are enzymatically phosphorylated. The process requires a "signal patch" of several amino acids brought into correct register by proper folding of the glycoprotein. Phosphorylation of mannose residues [mannose-6-phosphate (M-6-P)] at this stage protects them from removal by mannosidases in the Golgi cisternae.

M-6-P receptors in the TGN membrane bind M-6-P-labeled glycoproteins and concentrate them in forming transport vesicles.

This process is similar to receptor-mediated endocytosis occurring in clathrin-coated pits of the plasma membrane. The transport vesicles containing M-6-P-labeled acid hydrolases bud from the TGN, lose their clathrin coats, and are transported along microtubules to fuse with late endosomes.

---

pH plays an important role in the delivery and function of lysosomal hydrolases.

An acidic pH is maintained in late endosomes and lysosomes by proton pumps in the membranes of these organelles. The acidic environment of the late endosomes releases the M-6-P-labeled hydrolases from the M-6-P receptors. In addition, the hydrolases, which are inactive in the near-neutral pH of the TGN and transport vesicles, become activated in the acidic environment of the lysosome.

Membrane and unoccupied M-6-P receptors are recycled to the TGN by transport vesicles.

Lysosomal phosphatases remove the phosphate from the mannose residues of the lysosomal hydrolases, preventing their reattachment to M-6-P receptors.

M-6-P receptors function as bounty hunters and skip tracers.

Under normal conditions, a small quantity of M-6-P-labeled proteins escape to the cell surface via the constitutive (default) secretory pathway. Membrane-bound M-6-P receptors also reach the cell surface and retrieve the escaping proteins. This process is facilitated by the high affinity of M-6-P receptors for M-6-P-labeled proteins at the near neutral pH of the extracellular environment. The recaptured hydrolases are returned to lysosomes via receptor-mediated endocytosis.

I-cell disease is an example of pathology resulting from defects in the transport of lysosomal enzymes from the ER to lysosomes.

In I- (inclusion) cell disease, defects in a specific enzyme, *N*-acetylglu-cosamine-1-phosphotransferase, result in failure to properly label proteins for transport to lysosomes with mannose-6-phosphate. Several enzymes are therefore not transported to lysosomes, but are secreted into the blood. Undigested material (inclusions) may accumulate in the lysosomes. Defects in synthesis of other enzymes destined for lysosomal transport also result in accumulation of lysosomal debris and may disrupt function of some organs more than others.

TRANSPORT TO THE NUCLEUS

Proper function of the nucleus depends on its ability to (1) import from the cytoplasm large proteins, such as the histones and nucleoplasmin, required for the proper packing of the DNA and (2) export ribosomal subunits, tRNA, and mRNA for cytoplasmic protein synthesis. The double nuclear membrane facilitates separation of the nuclear and cytoplasmic compartments, allowing selective transport of some proteins and allowing exclusion from the nucleus of potentially deleterious proteins such as RNases and DNases.

Nuclear pores allow the nonselective transport of ions and small molecules (e.g., simple sugars) and regulate the import of targeted proteins.

The structure of nuclear pores has previously been described (see Chap. 4). Filamentous components of the nuclear pores serve as recognition or binding sites for molecules to be transported in either direction through the pore complex.

Proteins destined for import from the cytoplasm into the nucleus contain a nuclear localization signal.

In contrast to the terminal signal sequences routing proteins to the ER and mitochondria, the *nuclear localization signal (NLS)* may occur anywhere in the amino acid sequence of proteins targeted to the nucleus. The NLS consists of a *short stretch of amino acids (four to eight) rich in positively charged residues.* The only requirement is that *the NLS must be exposed on the surface of the folded protein for import to proceed.* Filamentous elements of the nuclear pore complex, with the aid of soluble cytoplasmic proteins, recognize the signal sequence and position the import protein in the core of the complex. Active transport of the import protein into the nucleus occurs via a mechanism requiring the *hydrolysis of ATP.* (Think of a giant octopus grabbing with its sucker-lined tentacles, positioning someone in front of its gaping maw and then shoving that person down its greasy gullet).

Nuclear localization signals are not excised after import.

There may be several reasons why the NLS is not removed from a protein after it is imported into the nucleus. The most obvious is that retention of the signal allows the protein to be recycled after mitosis. During mitosis, the nuclear membrane breaks down and the nuclear contents are dispersed throughout the cytoplasm. After cell division, nuclei reform in both daugther cells. Since the proteins targeted for nuclear import still retain their signal sequences, they can be reused by the daughter nuclei, thus reducing the need for new protein synthesis. Other differences between nuclear transport and the transport of proteins across other membranes include the use of an aqueous pore versus a membrane-bound protein transporter and the requirement for proteins to be in the folded versus partially unfolded state.

TRANSPORT TO MITOCHONDRIA

Mitochondria contain their own DNA, RNA, and ribosomes. However, the few proteins synthesized "in house" are only a small fraction of those which must be imported from the outside world, that is, the cytoplasm.

Proteins for distribution to mitochondria are synthesized on free cytoplasmic ribosomes and directed to the mitochondria by N-terminal signal sequences.

The mitochondria-directing signal of these proteins is generally longer and more complex than the signal sequence directing nascent proteins to the ER. Basically, the signal sequence consists of a *terminal region rich in positively charged amino acids* followed by a domain of alternating polar and nonpolar residues (*amphipathic domain*). This may be followed by another signal sequence which directs routing of the protein to a specific mitochondrial compartment. Because of their unique structure, mitochondria contain four possible destinations for targeted proteins: (1) the outer membrane, (2) the intermembrane space, (3) the inner membrane, and (4) the matrix compartment.

> Proteins are imported into the matrix space at sites of close contact between the outer and inner mitochondrial membranes.

For proteins to be incorporated into mitochondria, they must be *maintained in a partially unfolded state* by a molecular chaperone of the *heat shock protein 70* (hsp 70) family. Once the signal sequence binds to a receptor on the surface of the outer membrane, the amphipathic segment penetrates the membrane and the remainder of the protein is transported across the two membranes into the matrix where the first signal sequence is cleaved. *Both the transport step and the maintenance of the unfolded state require ATP.* If present, a second signal sequence may direct the protein to the inner membrane or the intermembrane space. Some proteins directed to the outer membrane appear not to require the standard mitochondrial signal sequence.

CELL-CELL, CELL-MATRIX INTERACTIONS

·

- **Overview of Cell-Cell Interactions**
- **Components of the Junctional Complex**
- **Occluding Junctions**
- **Anchoring Junctions**
- **Communicating Junctions**
- **Cell-Cell Adhesion: Cell Adhesion Molecules**

· · · · · · · · · · · ·

With the major exception of blood and lymphoid cells, most cells do not float freely in the body, but are linked to each other and to the surrounding environment (extracellular matrix, ECM). The ECM is secreted by the resident cells and is composed of supporting molecules (collagens and proteoglycans) as well as extracellular linkage molecules. The shape of cells, their capacity for motility, and their polarity are determined by how the cell interacts with its neighboring cells and the extracellular matrix. These interactions are known as cell-cell and cell-matrix interactions, respectively.

In Figs. 7-1 to 7-12 the small intestinal epithelium is used as an example of cell-cell and cell-matrix interactions. Cells in the small intestinal epithelium sit in a linear arrangement with apical surfaces facing the lumen, lateral surfaces linked to adjacent cells, and basal surfaces connected to the underlying basement membrane which the epithelium synthesizes (Fig. 7-1). Lateral surfaces are linked to neighboring cells by junctional complexes across their lateral borders and to the basal surface by ECM-ECM-receptor transmembrane linkages.

OVERVIEW OF CELL-CELL INTERACTIONS
Junctional complexes

Why do cells need to be linked to one another if they are linked to the basement membrane? In terms of cell neighbors, like cell types hang with their homies, whereas unlike cells are excluded. This is not atypical of human relationships as well (despite the commentary that opposites attract, this often makes for short-duration marriages). Cell recognition and arrangement specificity are important in terms of sorting during development and establishment of tissues with a distinct cellular organization, like the epithelia which line organs with lumens (lumena) and surfaces of the body. The junctional complexes serve a number of key functions.

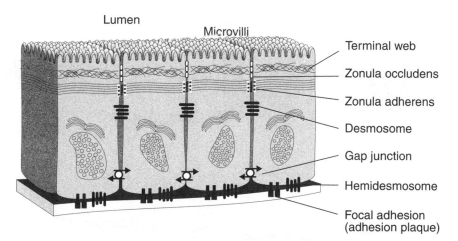

Figure 7-1 Epithelial cells on basement membrane showing location of junctional complexes laterally and links to basement membrane (integrins) basally.
The zonula occludens, zonula adherens, desmosome, and gap junction form the junctional complex. Focal adhesions (adhesion plaques) and hemidesmosomes attach the cells to the basement membrane.

Junctional complexes serve as a barrier to flow from the lumen to the intercellular space and vice versa (Fig. 7-2).

In the case of the small intestinal epithelium, the cells sit in a gunk of
digested foodstuff and enzymes. This material is absorbed by the cells, but the
intercellular spaces must be protected from luminal leakage. Also, material in the
intercellular space is blocked from passage to the lumen. This prohibits transport
of molecules from the intercellular space to the lumen.

Junctional complexes maintain apical versus basolateral polarity within the
cell.

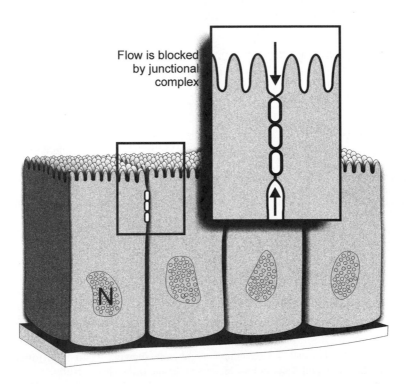

Figure 7-2
Epithelial cells illustrating lumenal surface and function of junctional complex as an
obstruction to flow from the lumen and also between the cells toward the lumen.

Most cells have a distinct polarity in which different molecules are sorted to the apical versus basolateral membranes (Fig. 7-3) in keeping with the diverse functions of these surfaces. When junctional complexes are broken down and cells are disaggregated, they lose their ability to maintain their apical versus basolateral polarity (see Fig. 12-1).

Junctional complexes help maintain cell shape.

Some junctional complexes are linked to the cytoskeleton or internal scaffolding of the cell. The junctional complex, along with the basement membrane, promotes cell rigidity and helps the cell resist flattening and shearing forces (Fig. 7-4).

Junctional complexes provide a means of cell-cell communication (Fig. 7-5).

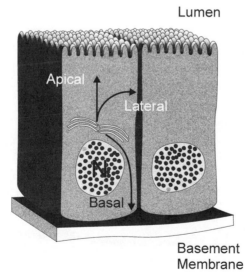

Lumen

Apical

Lateral

N

Basal

Basement
Membrane

Figure 7-3 Epithelial cells illustrating epithelial polarity.
Proteins are directed specifically to apical and basolateral membranes from the Golgi apparatus (located in a perinuclear location). The apical-basolateral domains are maintained by the presence of the junctional complexes.

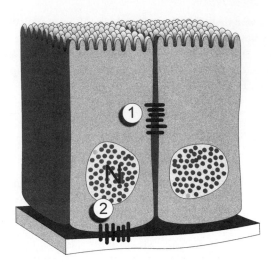

Figure 7-4
Epithelial cells illustrating the role of junctional complexes [at lateral cell borders, (1)]
and integrin-binding [basement membrane at basal surface (2)] in maintaining cell shape
and rigidity.

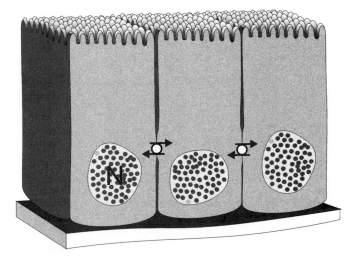

Figure 7-5
Epithelial cells illustrating the function of cellular communication, which is facilitated by
specific components of the junctional complex.

One of the specialized parts of the junctional complex is dedicated to cell-to-cell movement of molecules through a pore complex.

COMPONENTS OF THE JUNCTIONAL COMPLEX

The classic junctional complex consists of three parts: zonula occludens (tight junction), zonula adherens, and macula adherens (desmosome). The components differ in functional classification as well: occlusion (tight junction), adhesion (zonula adherens and desmosome), and communication (gap junction).

Components of the junctional complex involved in cell-cell communication are functionally and structurally similar to the structures involved in cell-matrix interactions.

Because of these similarities, cell-cell and cell-matrix structures are best described using a functional classification: 1) occluding junctions, 2) anchoring junctions, and 3) communicating junctions.

OCCLUDING JUNCTIONS
Zonula occludens, or tight junction

The zonula occludens provides (1) a barrier to diffusion between cells and (2) a barrier between apical and basolateral domains of epithelial cells.

The zonula occludens forms a continuous beltlike structure around epithelial cells. To achieve effective barrier function, the transmembrane proteins of the outer bilayers of the adjacent cells fuse to form sealing strands which eliminate the extracellular matrix and prevent leakage from the lumen. Backflow is also prevented by the tight junction. This is important in preventing molecules released by facilitated diffusion or exocytosis at the basolateral surface from seeping back into the lumen. The zonula occludentes (plural) maintain apical versus basolateral polarity. They can inhibit membrane fluidity and block the passage of ions and even water. Tight junctions are responsible for some of the selectiveness of the permeability barrier in that they contain ion-selective pores.

The tightness of a zonula occludens is related to the number of sealing strands present between the cells.

The sealing strands are visible in a freeze fracture preparation between adjacent cells. Figure 7-6 illustrates the beltlike arrangement of the tight junction.

ANCHORING JUNCTIONS

As the classification implies, these junctions anchor cells. They are characteristic of epithelial cell sheets and form a continuous link between the cells while strengthening the capacity of the sheet to resist mechanical stress.

Anchoring junctions are a broad classification of complexes which bind the cytoskeleton of adjoining cells to one another (cell-cell interaction) or to the ECM (cell-matrix interaction).

Anchoring junctions differ in the nature of the binding (cell-cell versus cell-ECM) and in the cellular splice: actin in the case of the adherent junctions and intermediate filament proteins in hemidesmosomes and desmosomes. There are four subclassifications of anchoring junctions based on these distinctions: zonula

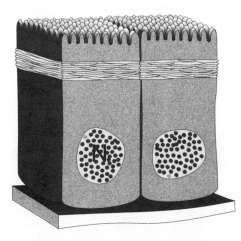

Figure 7-6
Epithelial cells illustrating the zonula occludens (tight junction) and arrangement of transmembrane proteins that actually encircle the entire cell as a belt.

adherens and desmosomes which bind in a cell-cell manner and focal adhesions and hemidesmosomes which bind cells to the ECM.

Zonula adherens

> The zonula adherens contains transmembrane proteins known as cadherins which link the actin microfilaments of one cell with actin in the neighboring cell.

The zonula adherens (Fig. 7-7), like the zonula occludens, encircles the entire cell. The *cadherins* are the key elements of this junctional complex and are Ca^{2+}-dependent molecules with distinct functional domains: adhesive recognition, calcium binding, transmembrane, and cytoskeletal interactivity. Each domain is important in the function of the cadherin molecule which recognizes cadherin from an adjacent cell, crosses the membrane, and interacts with intracellular attachment proteins of the cytoskeleton. These attachment proteins (*catenins*) serve as a bridge to bind actin. The binding of actin-cadherin-actin represents a homophilic-type interaction. It is important to note that the cadherins are also distributed throughout the plasma membrane of most cells in addition to their participation in the zonula adherentes (plural of zonula adherens).

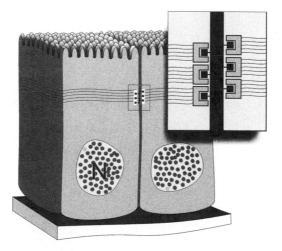

Figure 7-7
Epithelial cells illustrating the position and structure of the zonula adherens that forms a belt around the cells. It is linked to actin bundles by intracellular attachment proteins of the cytoskeleton, which also bind the cadherins.

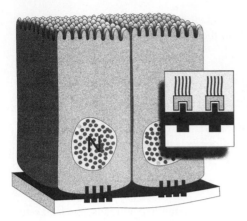

Figure 7-8
Epithelial cells illustrating the
structure and function of focal
adhesions that link actin indirectly (by intracellular attachment proteins) to the basement
membrane. The key ingredient
is the transmembrane linkers
(members of the integrin family).

Focal adhesions

Focal adhesions or contacts (Fig. 7-8) are similar to zonula adherens in that the actin cytoskeleton is the splice site. They differ in anchoring to the ECM rather than to neighboring cells and in the use of a different family of transmembrane proteins known as integrins in place of cadherins.

Integrins are glycoproteins which bind to numerous extracellular matrix molecules, among them collagens, fibronectin, and laminin. The integrins there-fore function as receptors for ECM molecules. The integrins tend to form tran-sient links between cells and the ECM which are broken down when cells move from that site. Integrin interactions are of the *heterophilic type* (e.g., fibronectin with fibronectin-receptor) in contrast to the *homophilic interaction* (e.g., actin-cadherin-actin of the zonula adherens). These interactions are discussed further in Chap. 8, "Cell Motility."

Macula adherens (desmosome)

The macula adherens, in contrast to the zonula occludens and zonula adherens, does not encircle the entire cell, but consists of rivets or spot welds.

The *desmosome* (Fig. 7-9) uses transmembrane linker proteins to connect the *intermediate filaments* of one cell with the intermediate filaments of the

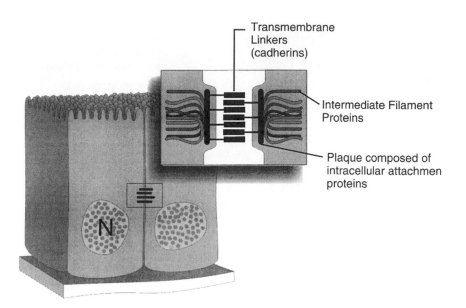

Figure 7-9
Epithelial cells illustrating the structure of the desmosome and its relationship to the cytoskeleton. Desmosomes are spot welds and do not encircle the cell.

adjacent cell. The filament type depends on intermediate filament-cell specificity (see Chap. 3, "Cytoskeleton"). The most obvious characteristic of the desmosome is the presence of a pair of dense plaques composed of intracellular attachment proteins known as the *desmoplakins*. The *desmocollins* fill the extracellular space between the plasma membranes of the adjacent cells. A third group of proteins, the *desmogleins*, forms a bridge between the ECM and the plaques.

Hemidesmosome

Although the name implies that a hemidesmosome (Fig. 7-10) is half of a desmosome, there are some key differences. First, the hemidesmosome contains different proteins from the desmosome. Second, the hemidesmosome links cells through their intermediate filaments to the ECM, whereas desmosomes are part of the junctional complex between cells and join the intermediate filaments of one cell with those of the adjacent cell. The hemidesmosome therefore represents a heterophilic interaction as compared with the homophilic interaction found in the desmosome.

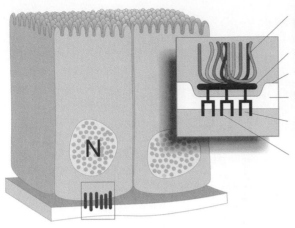

Intermediate filament proteins

Intracellular attachment protein

Cell membrane

Basement membrane

Binding to basal lamina proteins

Integrin

Figure 7-10
Epithelial cells illustrating the structure of the hemidesmosome. Transmembrane linkage occurs by integrins with an overall structure similar to half a desmosome.

COMMUNICATING JUNCTIONS
Gap junction (nexus)

The *gap junction* differs from the other parts of the junctional complex in that it functions primarily in communication and molecular transport between cells rather than exclusion of molecules (tight junctions) or adherence between cells. Gap junctions, as the name implies, form gaps between adjacent cells. In freeze fracture, gap junctions appear as a patch of several hundred specific channel-forming intramembranous particles called *connexons* (Fig. 7-11). Each connexon particle consists of six *connexins*. The connexins are transmembrane, multipass proteins which form a hexagonal arrangement surrounding a central channel pore. The connexins are tissue-specific. The connexon of one cell aligns with the connexon of an adjacent cell forming an *aqueous pore* or channel through which molecules of 1500 Da or less can move between the cells.

Terminology

Light-microscopic terminology for the junctional complex is ambiguous. The term *terminal bar* is used to describe the light microscopic thickening which is observed along the lateral membranes between cells. Technically, it represents the light microscopic view of the entire junctional complex. The terminal web is a network of actin microfilaments in the apex of the cell.

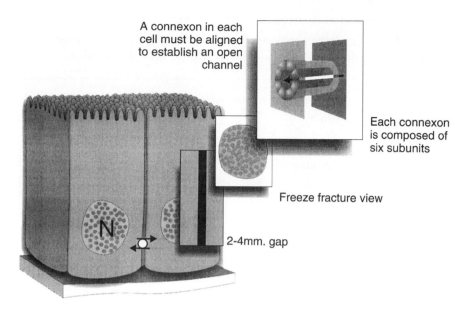

Figure 7-11
Epithelial cells illustrating the function and structure of the gap junction and its component connexons.

CELL-CELL ADHESION: CELL ADHESION MOLECULES

In addition to the junctional complexes which link cells together, there are membrane properties which regulate adhesion.

Like cells aggregate and form clusters, whereas unlike cells tend to disaggregate or be excluded from clusters. This characteristic is mediated by cell adhesion molecules (CAMs).

Although adhesion molecules may be involved in the formation of junctional complexes, they also may be found distributed throughout the plasma membrane. These molecules are responsible for the selective cell-cell adhesion and reassembly which occurs when cells are disaggregated. They provide the explanation for the affinity of like cells for one another.

Cell adhesion molecules may be classified into two categories based on calcium-dependency: (1) Ca^{2+}-dependent and (2) Ca^{2+}-independent.

Removal of calcium from an *in vitro* system results in the disruption of calcium-dependent cell adhesion.

The cadherins are one class of calcium-dependent cell adhesion molecules. They function through a homophilic mechanism.

Cadherins are single-pass transmembrane proteins with multiple Ca^{2+}-binding sites. They function as CAMs as well as constituents of the zonula adherens. Cadherins function through a homophilic mechanism in which a cadherin molecule on the surface of one cell recognizes a cadherin molecule on the cell surface of an adjacent cell. Cadherins also interact with the actin cortex of the cell through catenins, a group of intracellular attachment proteins which link the cadherins to the cytoskeleton. Cadherins have been identified in numerous tissues. Disaggregated cells with similar cadherins adhere to one another, whereas those with different cadherins tend to disaggregate.

The *selectins* are another class of calcium-dependent cell adhesion molecules. They function through carbohydrate binding.

The surface of cells is covered with a *glycocalyx* consisting of carbohydrates predominantly in the form of glycoproteins and glycolipids. Transmembrane proteoglycans and adhering carbohydrates also contribute to the carbohydrate content of the noncytoplasmic surface of the plasma membrane. Carbohydrate-binding proteins called *lectins* recognize specific sugar molecules on the cell surface. The *selectins* contain a lectin domain which recognizes and binds to specific sugar molecules on the surface of cells. Selectin function is usually transient and commonly involves the interaction of bloodborne cells with the lining cells of blood vessels. One of the best examples of selectin expression and function occurs during an inflammatory response where polymorphonuclear leukocytes (neutrophils) bind to the lining cells of blood vessels (endothelial cells). The selectin-mediated adhesion allows the neutrophils to roll over the endothelium where other interactions involving the extracellular matrix receptors (integrins) function to stabilize the interaction and "buy enough time" for the cells to break down intercellular junctions and facilitate tissue entry. The role of selectins is discussed in the chapters discussing the cardiovascular system (Chap. 18) and lymphoid tissues (Chap. 19).

> The calcium-independent adhesion molecules include the neural cell adhesion molecule (N-CAM), which is a member of the immunoglobulin superfamily.

There is a great deal of variety in the detailed structure and interactions of N-CAMs. However, it is clear that these molecules play a major role in development. One example is shown diagrammatically in Fig. 7-12. Neural crest cells adhere to one another (high N-CAM expression) until they are released from their

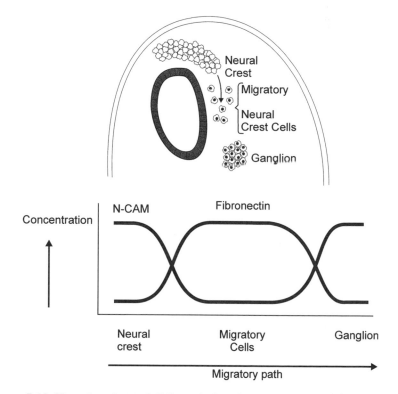

Figure 7-12 The migration and differentiation of the neural crest cells.
The migration and differentiation are regulated by extracellular matrix molecules (such as fibronectin), which pave the highway, or migratory pathway, for neural crest migration and eventual differentiation. N-CAM expression is critical to the aggregation-disaggregation events required for migration and establishment of ganglia.

adhesive bond and enter the migratory pathway (down-regulation of N-CAM). The neural crest cells leave the migratory pathway and aggregate to form the sympathetic chain ganglia. At their destination the neural crest cells differentiate into sympathetic ganglion cells and expression of N-CAM is maintained at high levels.

The complexity of cell-cell and cell-matrix interactions is reflected in the fact that more than one type of interaction may be involved in the same cell activity. For example, the adhesion, rolling, and entry of cells from the bloodstream into the tissues involves both selectin-mediated and integrin-mediated steps (see Chap. 18). Other players also exist which are important in determining the stickiness and adhesion between cells and their neighbors as well as the extracellular matrix. *Syndecan* is an example of a transmembrane proteoglycan which both binds the actin cytoskeleton and sticks to glycosaminoglycans (GAGs) across the cell membrane. This GAG-dominated extracellular domain binds extracellular matrix molecules as well as an array of growth factors. These factors can regulate cell proliferation, migration, and other cell functions.

CELL MOTILITY

·

- **Actin-based Motility in Nonmuscle Cells**
- **Actin and Myosin-based Contraction in Muscle Cells**
- **Sarcomere**
- **Contraction in Smooth Muscle Cells**
- **Microtubule-based Intracellular Movement**

·   ·   ·   ·   ·   ·   ·   ·   ·   ·   ·   ·

This chapter concentrates, for the most part, on two phenomena: (1) how cells move within their environment (nonmuscle cells) and (2) how cells move the environment around them (muscle cells). Both types of movement depend on the biochemical and physical properties of one protein, actin. Two functions of actin form the foundation of cell motility: (1) the ability of individual actin molecules (monomers) to reversibly polymerize into actin filaments and (2) the ability of actin to activate the energy-releasing ATPase activity of myosin.

How cells move their internal environment by both actin and microtubule-based systems is also considered. Microtubules polymerize from tubulin subunits in a similar fashion to actin filaments and share other similarities with the actin system including the use of molecular motors.

The biochemistry and dynamics of actin filaments and microtubules are discussed in Chap. 3, "Cytoskeleton."

ACTIN-BASED MOTILITY IN NONMUSCLE CELLS

Cells move by extending wide (lamellipodia) or narrow (microspike) processes induced by the polymerization of actin filaments.

Actin is concentrated in a layer around the periphery of the cytoplasm beneath the plasma membrane (*cortical actin*). Extension of a leading edge of the plasma membrane involves transport of new membrane components along microtubules and the polymerization of actin into growing filaments at a focal site.

G-actin is added to growing actin microfilaments at nucleation sites at the leading edge.

It is not yet clear whether the growing actin microfilaments are anchored to the plasma membrane and push the leading edge of the lamellipodium forward or whether they are quickly released to be added to a growing meshwork of filaments behind the leading edge.

Environmental signals (e.g., chemotactic factors) direct cell movement (e.g., chemotaxis) toward a specific target.

The cellular response involves cell-surface receptors which directly or indirectly regulate the polymerization of actin. Binding of a *chemotactic factor* to a receptor initiates a series of events which includes the activation of G proteins or receptor-linked enzymes (Chap. 9). For example, one result of binding of a chemotactic factor to its cell-surface receptor may be the *disinhibition* of actin polymerization mediated by actin-binding proteins such as profilin and actin destabilizing factor (ADF) (Chap. 3).

Cell movement requires the release of cells from focal adhesions to other cells or extracellular matrix components and the establishment of new contacts.

Transmembrane proteins of the cell adhesion molecule family, such as the *integrins*, link the actin cytoskeleton to extracellular matrix components (e.g., fibronectin) at sites of focal adhesion (Fig. 8-1).

Stress fibers form where tension is generated at cellular attachment sites between cells or between cells and the extracellular matrix (ECM). They counteract outward forces on the cell membrane.

Stress fibers contain actin filaments of mixed polarity as well as polymers of myosin II which serves as a molecular motor. Stress fibers are linked end-on at

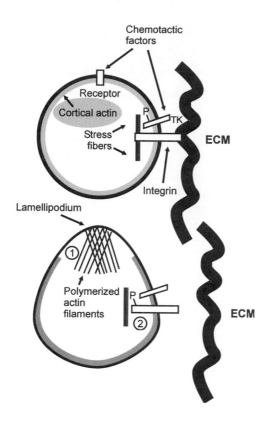

Figure 8-1 Roles of cortical actin and stress fibers in cell detachment and movement away from focal adhesions. Binding of chemotactic factors drives two processes in cellular migration: (1) polymerization of actin from actin monomers in the actin cortex drives the formation of lamellipodia in the direction of movement and (2) phosphorylation of integrins (e.g., the fibronectin receptor) by tyrosine kinase (TK)–linked receptors leads to dissociation of integrins from the actin cytoskeleton, allowing cells to detach from the extracellular matrix (ECM).

the plus ends of the actin filaments to integrins via accessory proteins α-*actinin, talin, and vinculin*. At their other ends, stress fibers may be linked to other focal adhesions or to a network of intermediate fibers surrounding the nucleus. Stress fibers are similar to the actin and myosin filaments associated with adhesion belts found in epithelial cells (Chap. 7).

Stress fibers are generally transitory structures.

When cells such as fibroblasts form focal adhesions to the ECM *in vivo* or to a substrate *in vitro*, stress fibers form at the focal adhesion sites. In cultured cells, particularly, this leads to flattening. Phosphorylation of the cytoplasmic portion of integrins by *tyrosine kinase-linked receptors* at the focal adhesion sites results in dissociation of integrins and other small attachment proteins from the actin cytoskeleton and allows the cell to move away from the adhesion site (Fig. 8-1). When the cells detach from the substrate, the actin filaments depolymerize, the tension on the plasma membrane is canceled, and the cells become rounded. A similar process occurs during mitosis.

> Actin and myosin filaments form the contractile ring during mitosis.

Actin filaments of mixed polarity begin to polymerize along with myosin II and α-actinin in the region of the cleavage furrow before telophase. (The phases and processes of mitosis and meiosis are reviewed in Chap. 10.) The motor function of myosin draws the actin filaments together like a purse string and separates the cytoplasm and daughter nuclei into new cells.

> Actin forms the core of microvilli.

Long, parallel arrays of actin filaments, bundled together by accessory proteins *villin and fimbrin*, are attached to the membrane walls of the microvilli by *myosin I and calmodulin*. The actin bundles are linked to each other and to intermediate filaments by myosin II and spectrin II (see Chap. 3). Despite the presence of actin and myosin, microvilli are generally considered nonmotile.

ACTIN AND MYOSIN-BASED CONTRACTION IN MUSCLE CELLS

Myosins are a family of proteins which, like some of the proteins discussed previously, bind to actin. They may be considered as molecular "motors" somewhat similar to dynein and kinesin (see below). Myosins are divided into nonmuscle myosins (myosin I) and muscle myosins (myosin II).

> Muscle myosin consists of two identical heavy chains, each divided into a long helical tail and a globular head.

The two helical portions wrap around each other to form a "coiled coil," whereas the head portions of heavy myosin chains remain separate. The effect is somewhat similar to the appearance of a caduceus (Fig. 8-2). Each head region is associated with two light chains, the essential light chain and the regulatory light chain.

Figure 8-2 The caduceus.
Muscle myosin resembles the
caduceus. The tails of the snakes
represent the helical portions of
the two myosin molecules intertwined in a coiled coil. The separate snake heads represent the
globular portions of the myosin
molecules.

> The myosin heavy chain head regions bind to actin and possess actin-
> stimulated ATPase activity.

The hydrolysis of ATP to ADP + P_i (orthophosphate) provides the energy
for the conformational changes driving muscle contraction. In the absence of
ATP, myosin remains tightly bound to actin, and movement of myosin along the
actin chain does not occur (*rigor*). The following stages characterize one cycle of
the "sliding filament" model of muscle contraction (Fig. 8-3):

1. **Resting phase:**   In the absence of ATP, myosin heads are tightly bound to
   actin.
2. **ATP binding:**   Binding of ATP to the myosin head results in separation of
   the myosin head from the actin filament.
3. **Conformational change:**   A large change in the shape of the myosin
   head displaces it *toward the plus end of the actin filament*. Coincidently,
   ATP is hydrolyzed to ADP + P_i, which remain tightly bound to the myosin
   head.
4. **Power stroke:**   The myosin head becomes tightly attached to a new site on
   the actin chain (closer to the plus end). The myosin head is now "cocked." A
   subsequent conformational change in the myosin head pulls the actin filament *toward the minus end* of the actin.
5. **Recovery phase:**   Release of ADP returns the myosin head to the "resting"
   stage.

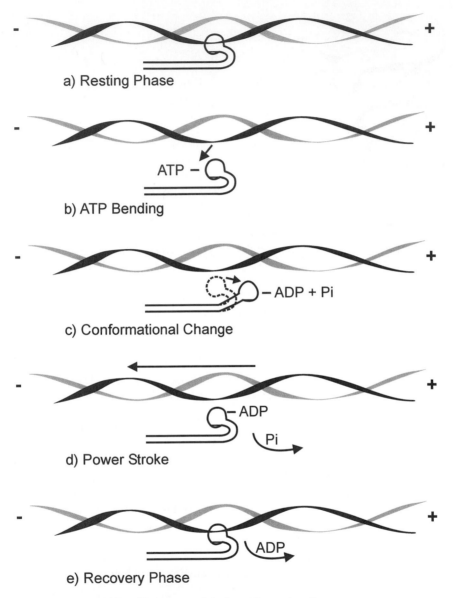

a) Resting Phase

ATP —

b) ATP Bending

— ADP + Pi

c) Conformational Change

— ADP

Pi

d) Power Stroke

ADP

e) Recovery Phase

Figure 8-3 The sliding filament model of muscle contraction.
The stages of muscle contraction are described in the text.

Intracellular Ca^{2+} concentration regulates muscle contraction.

In the striated muscles, depolarization of the cell membrane by neurotransmitters spreads into the interior of the cell via the *t-tubule system*. Elements of the t tubules come into close association with portions of the endoplasmic (sarcoplasmic) reticulum where Ca^{2+} is stored. Activation of channels in the membrane of the sarcoplasmic reticulum allows Ca^{2+} to "leak" into the surrounding cytoplasm.

The Ca^{2+}-dependence of muscle contraction is mediated through accessory proteins.

Tropomyosin is structurally similar to myosin and occupies the myosin-binding site on actin in resting muscle cells, preventing the binding of myosin. Increased intracellular Ca^{2+} stimulates the release of tropomyosin from actin, freeing actin to bind myosin. The effects of Ca^{2+} on tropomyosin are mediated through the *troponin complex*. The troponin complex consists of three protein subunits: *troponin T* binds tropomyosin and positions the troponin complex on the actin filament; *troponin C* is a calmodulin-like protein which binds Ca^{2+}; *troponin I* inhibits the actin-myosin interaction when Ca^{2+} is not bound to troponin C.

SARCOMERE

Actin and myosin form permanent assemblies in striated muscle cells. Individual units (sarcomeres) (Fig. 8-4) are arranged end-to-end to form myofibrils.

The Z disk anchors the actin filaments of the sarcomere.

The Z disk is found at either end of a sarcomere and consists of several proteins including α-actinin, which binds the plus ends of actin filaments. It is interesting to note that α-actinin is found in other structures where actin is anchored, including intercalated discs of cardiac muscle, dense bodies of smooth muscle, and focal contacts. *Nebulin* is a large protein with repeated actin-binding regions

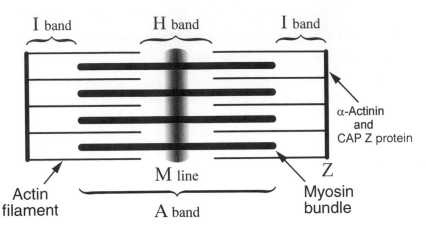

Figure 8-4 The sarcomere.
The Z disk (Z) at either end of the sarcomere contains several proteins including α-actinin, which serves as an attachment site for actin filaments. The I bands are regions consisting primarily of actin filaments, whereas the A band consists of overlapping actin filaments and myosin bundles. The H band is a region consisting primarily of myosin bundles, whereas the M line represents interlinkage between the myosin bundles.

which appears to form a scaffold for actin filament assembly and may regulate the length of the actin filaments.

Bundles of myosin filaments are arranged tail-end to tail-end in the center of the sarcomere.

The tail-to-tail arrangement of bundles of myosin filaments allows the active head regions to be aimed toward both Z disks (Fig. 8-5). Several proteins hold the

Figure 8-5 The asparagus model of sarcomere myosin bundles.
Myosin bundles in the center of the sarcomere are arranged end to end like these bunches of asparagus. In this model, the myosin heads (asparagus flowers) face the Z line at either end of the sarcomere, whereas the myosin tails (asparagus stalks) abut in the center.

myosin filaments in register in the center of the sarcomere. *Titin*, the largest protein yet discovered, anchors the myosin filaments to the Z disks, but is elastic enough to allow sliding of the myosin filaments with respect to actin.

Actin and myosin filaments overlap in the A band of a resting sarcomere.

A cross section through the center of the sarcomere (Fig. 8-6) reveals that the myosin filament bundles are positioned in a hexagonal array such that each filament bundle is surrounded by six other bundles.

A similar section through the I band would reveal only the thin actin filaments, also in hexagonal arrays (Fig. 8-7).

However, a section through the A band, where actin and myosin overlap (Fig. 8-8), would reveal that each myosin filament is surrounded by six actin filaments. In high-magnification electron micrographs, cross-bridges between the actin and myosin filaments, formed by the myosin head groups, can frequently be observed.

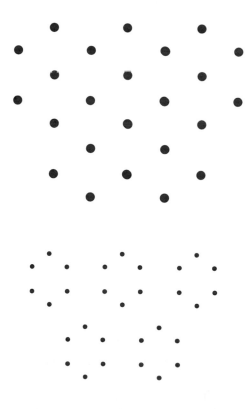

Figure 8-6 Cross section through the center (H band) of a sarcomere.
Only myosin bundles in a hexagonal array are present.

Figure 8-7 Cross section through an I band of a sarcomere.
Only actin filaments are present.

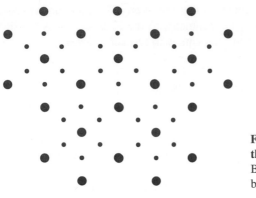

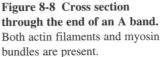

**Figure 8-8 Cross section
through the end of an A band.**
Both actin filaments and myosin
bundles are present.

During contraction, myosin filaments pull the actin filaments, and thus the
Z disks at either end, toward the center of the sarcomere.

Neither the actin nor the myosin filaments shorten during contraction. Only
the I band between the Z disk and the myosin filaments decreases in width during contraction (Fig. 8-9).

CONTRACTION IN SMOOTH MUSCLE CELLS

Typical sarcomeres are not present in smooth muscle cells.

In contrast to the sarcomeres of skeletal and cardiac muscle, parallel bundles
of actin and myosin filaments are distributed throughout the cytoplasm of smooth
muscle cells. Similarities do, however, exist. For example, actin filaments in
smooth muscle cells are anchored at their plus ends into cytoplasmic "dense bodies" which contain α-actinin. Therefore, dense bodies perform similar functions
to the Z disks of striated muscle (Fig. 8-10). Some dense bodies also anchor actin
filaments to the plasma membrane. Also, as in striated muscle, intermediate filaments (desmin and vimentin) stabilize the contractile apparatus and link it to the
plasma membrane.

Ca^{2+} regulates smooth-muscle contraction through a different mechanism
than in striated muscle.

The primary mechanism for regulation of contraction in smooth muscle cells
involves *Ca^{2+}-dependent phosphorylation of the regulatory light chain* associ-

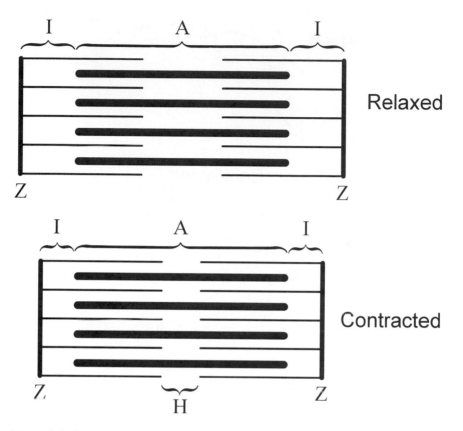

Figure 8-9 Contraction of a sarcomere.
The myosin bundles pull the actin filaments toward the center, resulting in shortening of the I bands.

ated with the myosin heads. An increase in intracellular Ca^{2+} concentration (primarily of extracellular origin) activates *calmodulin* which, in turn, activates an enzyme, *myosin light chain (MLC) kinase*. MLC kinase then phosphorylates the *regulatory myosin light chain*, allowing the myosin head to bind actin.

In smooth muscle cells, the troponin complex is replaced by lower concentrations of another Ca^{2+}-binding protein, *caldesmon*. Therefore, the actin-based mechanism of Ca^{2+}-regulated contraction plays a lesser role in smooth muscle than the myosin-based mechanism (MLC kinase).

> In contrast to skeletal muscle, contraction in smooth muscle may be inhibited (relaxed), as well as stimulated, by hormones and neurotransmitters.

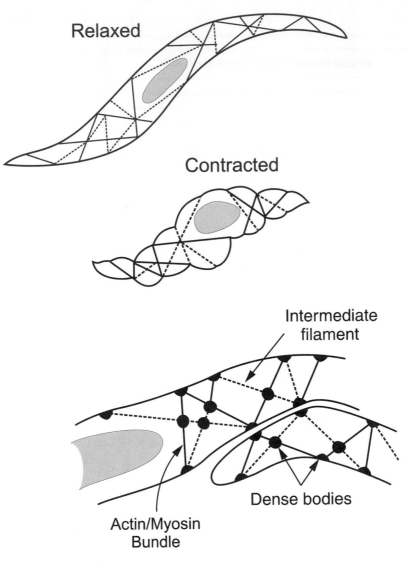

Figure 8-10 Contraction in smooth muscle cells.
Bundles of actin combined with myosin are distributed throughout the cell and attached to the cell membrane by dense bodies. Activation of the myosin heads draws the actin filaments centrally and results in "scrunching" of the smooth muscle cell.

Typically, skeletal muscle cells are stimulated to contract by acetylcholine released at neuromuscular junctions. The contraction is *rapid and of short duration*. In smooth muscle, however, the contractions stimulated by such agents as adrenaline and angiotensin II are *slow and of longer duration*. Other hormones or neurotransmitters such as β-adrenergic agonists relax smooth muscle by binding to surface receptors linked to G proteins. This eventually results in the phosphorylation of MLC kinase, which inhibits its ability to activate the regulatory myosin light chain.

MICROTUBULE-BASED INTRACELLULAR MOVEMENT

Kinesins and dyneins are "motor" proteins associated with microtubules.

Some members of the *kinesin* and *dynein* families of proteins share structural and functional similarities with the myosins; they are composed primarily of two heavy chains with globular heads and rodlike tails. The head regions possess ATPase activity and bind to microtubules. The tail regions bind to cellular components and organelles with specificity. Light chains are also associated with the heavy chains.

In most cases, the movement of motor proteins along microtubules is unidirectional.

Kinesins move toward the plus ends of microtubules and away from the *microtubule organizing centers (MTOCs)* of the cell. For example, they move organelles such as mitochondria, and especially synaptic vesicles, toward the distal (plus) ends of neuronal axons. Dyneins, on the other hand, move toward the minus ends of microtubules and toward the MTOCs. This property has proved extremely useful in *retrograde tracing* experiments where chemical markers injected in muscles, glands, and regions of the brain are taken up by axon terminals. The chemical markers are then transported via dynein along the microtubules to the cell bodies of neurons innervating the injected structures. In similar fashion, *anterograde tracing* (from cell body to axon terminal) makes use of kinesin transport. Transport is critical in neurons because of the long distances between the cell body and the termination of axons and dendrites. The rate of transport is critical in determining the rate of repair following neuronal injury.

The movements of cilia and flagella are based on microtubules and specific molecular motors.

Cilia and flagella are constructed as nine modified pairs of microtubules (one complete and one incomplete) arranged concentrically around a central pair of microtubules (9 + 2 configuration). This structure forms the core, or *axoneme*, and is ensheathed in plasma membrane (Fig. 8-11).

Dynein produces a bending movement in cilia and flagella.

Arms of ciliary dynein (similar to but not identical with cytoplasmic dynein) are attached to the complete (A) microtubule of each of the nine outer microtubule pairs (Fig. 8-11). *Nexin* connects the A and B tubules of adjoining doublets, and radial spokes connect the peripheral doublets to the central pair of microtubules. During ciliary or flagellar movement, the ends of the dynein arms contact the incomplete (B) tubule of an adjoining pair and move toward its minus end. Between "free" microtubule pairs, this movement would produce a sliding movement of one pair along the other. However, in cilia and flagella, the movement of adjacent microtubules is constrained by a variety of cross-linking proteins, resulting in a bending of the axoneme and a whip- or wave-like movement of the cilia or flagella (Fig. 8-12). The beating of many cilia propels some unicellular organisms through their aqueous environment. The cilia of stationary cells such as those lining the oviduct create movement in the aqueous environment which propels ova toward the uterus.

The activities of cilia and flagella are regulated by phosphorylation or dephosphorylation of associated proteins.

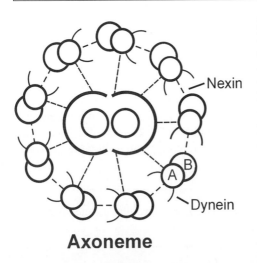

Axoneme

Figure 8-11 The axoneme.
Nine pairs (doublets) of microtubules surround a central pair. Each outer doublet consists of a complete (A) tubule and an incomplete (B) tubule. Each A tubule possesses a pair of dynein arms and is connected by the protein, nexin, to the B tubule of an adjoining doublet. Dynein is the molecular motor which slides each doublet along an adjoining one. The outer doublets are connected to the sheath surrounding the central pair by radial protein spokes. The axoneme forms the mechanical core of cilia and flagella.

There may be as many as 200 or more polypeptides associated with the axonemal microtubules of flagella. Although most of these have not been isolated (and therefore, fortunately, they don't have to be memorized), it is apparent that phosphorylation of some of these polypeptides, as well as portions of the microtubules, by a cyclic adenosine monophosphate (*cAMP*)-*dependent protein kinase* may serve as a molecular "on" switch to activate normally quiescent flagella. Conversely, dephosphorylation by *phosphorylases* acts as an "off" switch. Therefore, as in many other systems, phosphorylation is the driving force in the regulation of ciliary and flagellar function.

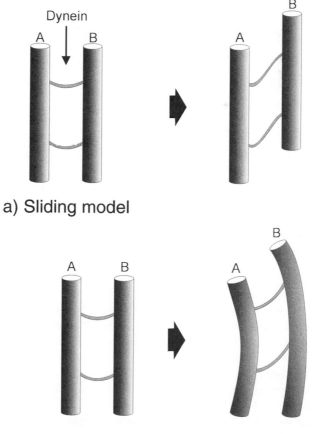

a) Sliding model

b) Bending model

Figure 8-12 Ciliary bending.
The action of dynein in a pair of unrestrained microtubules would result in the sliding of one past the other (*A*). However, in a cilium or flagellum, microtubule pairs are highly cross-linked and restrained in their freedom of movement. The action of dynein results in a bending, or "whip-like," movement rather than sliding (*B*).

CELL RECEPTORS AND INTRACELLULAR SIGNALING

·

- Receptor–G-Protein–Effector Complexes

- Activation of G-Protein Complexes

- General Principles

- Enzyme-Linked Receptors

- Steroids and Other Lipid-Soluble Compounds

- Gasses as Messengers

If a tree falls in the forest and there is no one around to hear it, does it make a sound? This old Zen aphorism is applicable to the realm of intracellular signaling. With the exception of the steroid hormones and the gaseous messengers (nitric oxide, NO; carbon monoxide, CO), both of which diffuse readily through the plasma membrane, communication between cells depends on receptors localized on the exterior of the cell membrane. However, these docking sites for chemical messengers (hormones and neurotransmitters, for example) are mute without a means of transducing the extracellular message to an intracellular one capable of initiating a series of events leading to the intended results (e.g., turning on or turning off a specific target cell activity). This chapter reviews two primary mechanisms (receptor–G-protein–effector complexes and enzyme-linked receptors) by which external signals are converted to intracellular "second" messages. Actions of the steroids and gasses are also briefly reviewed.

RECEPTOR–G-PROTEIN–EFFECTOR COMPLEXES

G proteins are a transducing system which link the binding of chemical signals (ligands) at extracellular receptors to the production of intracellular second messengers by effectors.

All receptors associated with G proteins are seven-pass transmembrane proteins.

The extracellular NH_2 end of the receptor is where the primary message circulating in the blood or interstitial fluid binds. The COOH end inside the cell, along with the third intracellular loop, is the part of the receptor which interacts with the G protein.

Remember that the portion of the receptor protein within the plane of the membrane must consist primarily of hydrophobic amino acids. Therefore, a person shown the structure of a membrane protein with seven hydrophobic amino acid sequences might suspect that it is a G-protein–associated receptor.

G proteins are heterotrimeric integral membrane proteins.

What this mouthful of a sentence really means is that G proteins consist of three different subunits which float in the plasma membrane like a big upside-down iceberg. *G proteins are situated on the inner surface of the plasma membrane.* They consist of an α subunit and a β/γ subunit pair (dimer) (Fig. 9-1). *The α subunit is variable and is specific for a particular G-protein complex.* More than 20 different G proteins have been characterized, based on their α subunit.

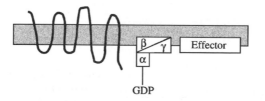

Figure 9-1 The receptor–G protein–effector complex.
The G protein–linked receptor is a seven-pass transmembrane protein. The G protein consists of α and β/γ subunits. The effector may be any number of enzymes such as adenylyl cyclase.

The β/γ subunits, while possessing some variability, are similar in most G proteins. The five best characterized G proteins (the only ones worth taking the trouble to know) are listed in the accompanying table.

CHARACTERIZATION OF G PROTEINS

G PROTEIN	RELATED ACTION AND EFFECTORS
G_s	Stimulates activity of adenylyl cyclase. [Increases intracellular level of cyclic adenosine monophosphate (cAMP).]
G_i	Inhibits activity of adenylyl cyclase. (Decreases intracellular level of cAMP.)
$G_{t/g}$	Transducin/gustducin stimulates breakdown of cGMP in photoreceptors and taste receptors, respectively. (Decreases intracellular cGMP levels by activating cGMP phosphodiesterase.)
G_o	Opens neuronal calcium channels. (Increases intracellular Ca^{2+} levels)
G_q	Activates phospholipases A and C. [Increases intracellular levels of inositol triphosphate (IP_3), which releases Ca^{2+} from intracellular stores].

The α subunit binds guanine nucleotides (GDP, GTP) with high specificity and affinity.

In the ground state (before binding of the ligand to the receptor), the α subunit binds GDP (Fig. 9-1). When activated, the α subunit exchanges GDP for GTP (Fig. 9-2). *Note:* The GDP is not converted to GTP, *it is swapped for a different molecule.* The resulting α-GTP is now able to activate the effector.

Effectors include a wide variety of agents which share no general functional or structural motifs (see table above).

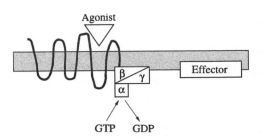

Figure 9-2 Activation of the G protein by a ligand.
When a ligand (agonist) binds to the receptor, the receptor undergoes a conformational change and interacts with the G protein, resulting in the exchange of GTP for GDP bound to the α subunit.

Effectors associated with G proteins include transmembrane proteins such as adenylyl cyclase, phospholipases, and ion channels as well as peripheral membrane proteins such as cyclic guanine monophosphate (cGMP) phosphodiesterase (Chap. 1).

ACTIVATION OF G-PROTEIN COMPLEXES

Binding of the ligand to the receptor activates the G-protein complex.

Binding of the ligand (Fig. 9-2) alters the conformation of the receptor and promotes interaction with the G protein heterotrimer, resulting in the following:

1.   Substitution of GTP for GDP on the α subunit
2.   Dissociation of the "activated" GTP–α subunit from the heterotrimer (Fig. 9-3)

The free GTP–α subunit stimulates or inhibits activity of a specific effector.

It is important to recall that these interactions (receptor–G protein and GTP–α subunit–effector) are occurring on the inner surface of the plasma membrane. As the activated receptor "iceberg" floats freely in the lipid sea of the plasma membrane, it is able to interact with multiple G-protein complexes resulting in *signal amplification*. (Visualize an old-fashioned pinball machine with the little steel ball bouncing back and forth between the plastic bumpers and adding to the points.)

Activation of the GTP–α subunit is self-limiting. The α subunit possesses intrinsic GTPase activity.

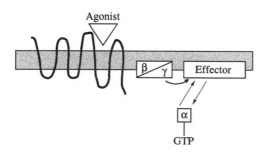

Figure 9-3 Dissociation of the activated α subunit.
The α subunit–GTP dissociates from the β/γ subunits and is free to interact with the effector. The β/γ subunits may also interact with the effector.

Once the system is turned on, there must be a mechanism for turning it "off". Otherwise, the system would remain on, resulting in overstimulation and loss of the ability to respond to new signals (much like the effect of 15 cups of coffee the night before an important final examination). The intrinsic GTPase activity of the α subunit converts GTP to GDP (this time it *is* a conversion and *not* an exchange) (Fig. 9-4). This inhibits the interaction of the α subunit with the effector and leads to reformation of the inactive G-protein heterotrimer. In fact, the effector protects itself from overstimulation by stimulating the GTPase activity of the α subunit. Indeed, some effectors are so good at this that they are classified as GTPase-activating proteins (GAPs). It is important to note that problems with the GTPase activity of the α subunit resulting either from genetic mutations or the actions of toxins (cholera, pertussis) can lead to extended, or even irreversible, activation of the α subunit and the effector.

Prolonged exposure to the ligand results in desensitization of the receptor. Phosphorylation of the receptor inhibits its association with the G-protein complex.

As learned from biochemistry, kinases are enzymes which add phosphate groups to other proteins. Receptor kinases are stimulated by the β/γ subunit (finally, something for these guys to do although they may also have some influence on certain effectors) to phosphorylate serine and threonine residues on the intracellular (carboxy) end of the receptor. However, the kinases find this difficult to accomplish effectively while floating freely in the cytoplasm. They need to be where the action is, that is, on the inner surface of the plasma membrane.

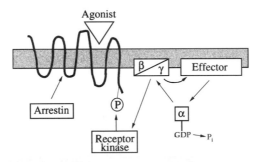

Figure 9-4 Deactivation of the receptor–G protein complex.
Interaction of the activated α subunit with the effector stimulates activity of an intrinsic GTPase, which returns the α subunit to the inactive state and favors reformation of the intact G protein. In addition, the β/γ subunit stimulates receptor kinases, which phosphorylate the receptor and inhibit its interaction with the G protein. Phosphorylation also promotes binding of arrestins, which further inhibit the receptor–G protein interaction.

To facilitate this change of address, they borrow a hydrophobic group from the β/γ subunit and use it to anchor themselves to the inner face of the plasma membrane. Phosphorylation of the receptor results in (Fig. 9-4) the following:

1. Reduced coupling of the G-protein complex to the receptor.
2. Promotion of the binding of *arrestins* to the receptor. Arrestins are a class of proteins which sterically inhibit the receptor–G-protein interaction.

GENERAL PRINCIPLES

Many different classes of ligand may activate the same type of G protein.

TYPICAL LIGANDS AND THE G PROTEINS THEY ACTIVATE

G PROTEIN	ACTIVATING LIGANDS
G_s	ACTH
	Adrenaline/noradrenaline (β receptors)
	Glucagon
	Luteinizing hormone
	Odorants
	Vasopressin
G_i	Adrenaline/noradrenaline (α_2 receptors)
G_q	Acetylcholine
	Angiotensin
	Thrombin
	Vasopressin

As can be seen from the incomplete lists in the table, some ligands (e.g., angiotensin) may activate only one type of G protein. Others (e.g., vasopressin) may activate different G proteins in different cells. Finally, some ligands (adrenaline/noradrenaline) may have opposing effects in different cells. *The differential effect of the ligands resides in the type of receptor expressed by particular cell types.*

G proteins control effectors, which eventually alter the intracellular concentrations of "second messengers."

Prominent among the second messengers are cAMP (G_s and G_i) and diacylglycerol (DAG; G_q) which activate protein kinases A and C, respectively. These enzymes, in turn, phosphorylate one or more other proteins leading to phosphorylation of a very specific protein (site-specific transcription protein) which can bind to DNA and regulate the activity of one or more specific genes. On the other hand, inositol triphosphate (IP_3; G_q) is another second messenger which releases Ca^{2+} stored in the smooth endoplasmic reticulum. The resulting increase in intracellular calcium concentration stimulates calcium-dependent kinases which add to the cascade of phosphorylation leading to activated transcription proteins. *Phosphorylation of regulatory proteins is the final common pathway in these systems.*

Summary

G proteins are one mechanism by which cells respond to external stimuli. Cell surface events such as ligand binding are thus translated into cellular responses such as stimulated secretion or cell proliferation using G proteins as intermediaries between extracellular and intracellular messages. In essence, each G protein is a black box which translates many separate, otherwise indistinguishable message "languages" into a common message which cells can understand and act on.

ENZYME-LINKED RECEPTORS

Enzyme-linked receptors are single-pass transmembrane proteins.

This class of independent-minded receptors does not rely on the aid of intermediary agents like G proteins. Members of this group contain or closely associate with an effector capable of generating an intracellular signal. Like the receptors associated with G proteins, they possess an extracellular portion which binds ligands (Fig. 9-5). Binding of the ligand activates an enzyme or catalytic region which is an integral part of the intracellular region of the receptor or is noncovalently bound to it.

| Ligand-binding | Transmembrane | Catalytic |
| domain | domain | domain |

Figure 9-5 General structural scheme of enzyme-linked receptors.
These receptors consist of a single transmembrane domain, an extracellular ligand-binding domain, and an intracellular catalytic domain.

The enzyme-linked receptors fall into two classes: (1) those which possess guanylyl cyclase activity and therefore increase intracellular cGMP concentration when activated; the receptors for the natriuretic peptides (ANP, BNP, CNP) are the best characterized of this group and (2) receptors which catalyze the phosphorylation or dephosphorylation of intracellular signaling proteins.

Activation of guanylyl cyclase–containing receptors increases intracellular cGMP.

There are three subclasses of receptors with intrinsic guanylyl cyclase (GC) activity (A, B, and C). GC-A and GC-B bind different members of the natriuretic peptide family with different specificities and affinities. These peptide hormones, derived primarily from the heart and central nervous system, lower blood pressure by increasing urinary excretion of sodium and water and by dilating most vascular beds. They also inhibit the synthesis of other hormones (aldosterone, angiotensin II, and vasopressin) which increase blood pressure. Therefore, the GC-A and GC-B receptors enjoy a wide distribution throughout the body. The third member of the group, GC-C, appears to be specific for bacterial enterotoxin and guanylin, a peptide which may regulate intestinal fluid absorption.

As noted above, activation of GC receptors results in increased synthesis of cGMP, which subsequently binds to, and activates, a cGMP-dependent protein kinase. As should be suspected by now, this kinase goes on to phosphorylate a cascade of other proteins which eventually regulate the transcription of specific genes.

Receptor Kinases and Phosphatases

Receptor kinases and phosphatases make up a large group of single-pass transmembrane receptors with extracellular ligand-binding domains, transmembrane α-helical portions, and intracellular catalytic units. These include receptor tyrosine kinases, tyrosine kinase–associated receptors, receptor tyrosine phosphatases, and receptor serine/threonine kinases. They are related to other proteins with kinase or phosphatase activity, but which lack ligand binding and transmembrane segments and are therefore cytosolic.

• **RECEPTOR TYROSINE KINASES**   This subclass of enzyme-linked receptors includes receptors for many growth factors as well as the insulin and insulin-like growth factor receptors. The extracellular ligand-binding domain may be either cysteine rich or immunoglobulin-like (see table).

RECEPTOR TYROSINE KINASES

EXTRACELLULAR DOMAIN	CYSTEINE-RICH	IMMUNOGLOBULIN-LIKE
	Epidermal growth factor (EGF) Insulin Insulin-like growth factor (IGF-1)	Nerve growth factor (NGF) Platelet-derived growth factor (PDGF) Macrophage colony stimulating factor (M-CSF) Fibroblast growth factor (FGF) Vascular endothelial growth factor (VEGF)

Following binding of ligand, receptor tyrosine kinases form dimers which cross-phosphorylate each other.

Catalytic enzyme-linked receptors are gregarious self-starters. After binding of the ligand, two adjacent receptors in the plasma membrane come close together and cross-pollinate each other at tyrosine residues with phosphate groups donated by ATP. This is earthworm sex at the molecular level! The mechanism of pair bonding may involve either conformational changes in the receptor or cross-linking of two receptors by the ligand itself.

Autophosphorylation enhances tyrosine kinase activity.

As a result of autophosphorylation, many tyrosine residues on the intracellular portion of the receptor become phosphorylated. This allows a single activated receptor to phosphorylate a number of different target proteins (Fig. 9-6). Specificity for particular target proteins is encoded in the amino acid sequence bordering phosphorylated tyrosine residues on the receptor. These sequences are recognized by specific binding regions of the target protein. The phosphorylated target proteins then go on to serve as intracellular messengers, blithely spreading the gospel of phosphorylation to a cascade of other target proteins.

Ras proteins are targets for activation by receptor tyrosine kinases.

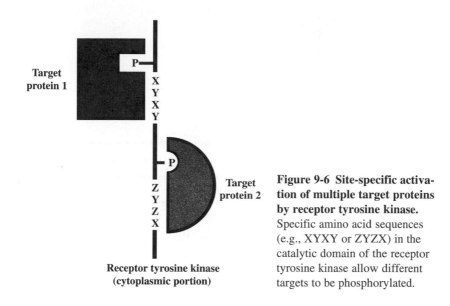

Figure 9-6 Site-specific activation of multiple target proteins by receptor tyrosine kinase. Specific amino acid sequences (e.g., XYXY or ZYZX) in the catalytic domain of the receptor tyrosine kinase allow different targets to be phosphorylated.

Ras proteins are part of a very large group of weak GTPases which control such diverse events as polymerization and organization of actin filaments as well as intracellular trafficking of transport vesicles. The ras proteins relay signals from activated receptor tyrosine kinases to a cascade of serine/threonine (nonreceptor) kinases which include the mitogen-activated protein (MAP) kinases (Fig. 9-7). In the simplest scenarios, activated MAP kinases migrate to the nucleus where they regulate the transcription of "immediate early" genes such as *fos* and *jun*. The protein products of *fos* and *jun* unite to regulate the activities of other genes involved in cell proliferation. It is important to note that MAP kinases form part of a final common signaling pathway which is accessible by both receptor tyrosine kinases and G-protein–linked receptors.

• TYROSINE KINASE-ASSOCIATED RECEPTORS

Tyrosine kinase–associated receptors bind extracellular messengers such as cytokines, growth hormone, and specific antigens.

Tyrosine kinase–associated receptors lack intracellular catalytic domains, but are intimately associated with nonreceptor protein tyrosine kinases. They comprise a large and varied group including the receptors for cytokines, growth hormone and prolactin, and the *antigen-specific receptors of lymphocytes*. The nonreceptor tyrosine kinases associated with these receptors are generally members of either the src or Janus families of proteins.

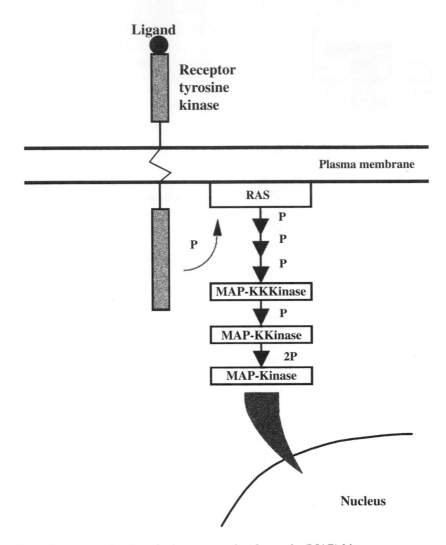

Figure 9-7 Ras activation of mitogen-associated protein (MAP) kinases.
Phosphorylation of ras protein by a receptor tyrosine kinase initiates a cascade of other phosphorylations including phosphorylation of a series of MAP kinases (MAP kinase kinase kinase → MAP kinase kinase → MAP kinase). MAP kinase regulates transcription of "immediate early" genes in the nucleus.

• RECEPTOR TYROSINE PHOSPHATASES

Receptor tyrosine phosphatases play an important role in reducing phosphorylated tyrosine to minimum levels in resting cells.

Members of this varied group of single-pass transmembrane proteins with intrinsic phosphatase activity in the intracellular catalytic domain dephosphorylate specific phosphorylated tyrosines on specific target proteins. This keeps the number of activated enzymes and signal proteins low when the cell is not stimulated by ligand.

• RECEPTOR SERINE/THREONINE KINASES

Transforming growth factor βs (TGFβ_{1-5}) bind and activate receptor serine/threonine kinases.

Many of the serine/threonine protein kinases are cytosolic and lack membrane-spanning and ligand-binding domains. This diverse group includes the well known kinases such as protein kinases A and C, calmodulin-dependent protein kinase, and MAP kinases. However, a class of single-pass transmembrane proteins with intrinsic serine/threonine kinase activity have recently been discovered which serve as receptors for the TGFβ family and related signal proteins such as the activins and bone morphogenetic proteins. These proteins play important roles in the regulation of vertebrate development.

STEROIDS AND OTHER LIPID-SOLUBLE COMPOUNDS

Steroid and thyroid hormones differ from the protein hormones in their mechanism of action on target cells. This difference is reflected in the type of cellular response: peptide hormones induce a response of short duration, whereas steroid/thyroid hormones induce a longer-lasting effect on the target cell.

Steroid hormones and thyroid hormones do not bind to cell surface receptors.

Because of their chemical nature, steroid and thyroid hormones are lipid soluble and are therefore able to easily penetrate the plasma membrane. Rather than binding to cell surface receptors, *they bind to intracellular receptors* (carrier proteins) in the cytoplasm or the nucleus.

Binding of steroid/thyroid hormones to their respective intracellular receptors exposes hidden DNA binding sites on the receptor.

Binding of these hormones to their receptors results in the formation of hormone-receptor complexes which will eventually bind directly to chromosomal DNA and activate transcription of specific genes (Fig. 9-8). The receptor proteins for both systems exhibit a similar structure: (1) an amino end involved in activation of the gene, (2) a DNA-binding domain, and (3) a hormone-binding domain at the carboxy end. The receptors exist in a folded, inactive state in the absence of hormone. However, binding of the hormone induces a conformational change in the receptor which releases an inhibitor protein and exposes the DNA-binding domain (Fig. 9-8).

GASSES AS MESSENGERS

Nitric oxide (NO) and carbon monoxide (CO) diffuse easily through the plasma membrane and activate the soluble (cytosolic, nonreceptor) form of guanylyl cyclase to increase intracellular cGMP content.

Until very recently, the idea of a gas as a messenger which could activate intracellular signaling was virtually unknown. Several investigators were working to identify a chemical messenger, endothelium-derived relaxing factor (EDRF), which relaxed vascular smooth muscle. Eventually, this substance turned out to be a gas (NO) and its generating enzyme, NO synthase, was discovered to be identical with a better known enzyme, NADPH diaphorase. Subsequent studies demonstrated the roles of NO as a neurotransmitter and cell toxin. NO also proved to be the active metabolite of nitroglycerin, a well known but poorly understood treatment for angina pectoris. Eventually, this shy little gas became bathed in the media spotlight because of its role in penile erection. Sex sells newspapers! Shortly after the discovery of NO, CO also became recognized as another gaseous messenger.

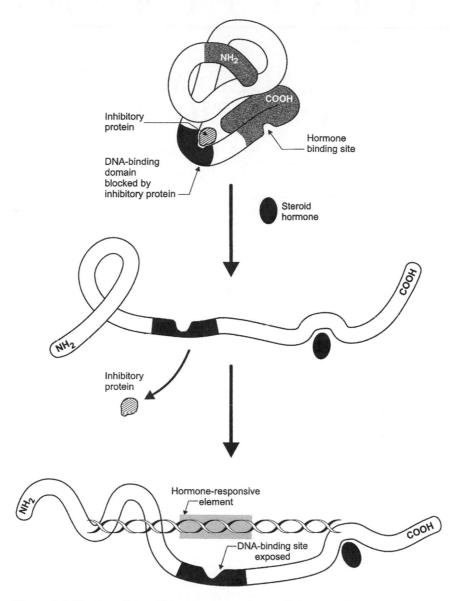

Figure 9-8 Binding of steroid hormone to an intracellular receptor.
Binding of steroid or thyroid hormones to their intracellular receptors exposes receptor DNA-binding domains. The receptors may then bind to DNA and regulate the transcription of specific genes.

CELL CYCLE: MITOSIS AND MEIOSIS

·

· · · · · · · · · · · ·

". . . they go round and round and the painted ponies go up and down, traveling on the carousel . . . in the circle game."

—Joni Mitchell

Dividing cells traverse a cycle of specific events with checkpoints to regulate entry into the next phase as the carousel turns round and round. As cells cycle they pass through a high and low *phosphorylation* state reminiscent of the up and down ponies on the carousel. The high phosphorylation state is instrumental in driving the key events of the cell cycle through phosphorylation of molecules associated with the nuclear envelope, the spindle assembly, and the structure of chromatin. The high and low phosphorylation state is the simplest way of viewing the cell cycle which, like the carousel, is regulated by a complex machinery of gears and motors representing subcycles within the cell cycle.

Cell division, or proliferation, is an essential event during development when the number of cells expands exponentially in each organ. In the adult, cell proliferation is much more limited. Organs which are classified as continuously renewing contain tissues which turn over and undergo cell proliferation throughout life. The epithelium of the gastrointestinal tract, the epithelium of the skin, and the progenitor cells of the bone marrow are examples of cell populations which provide for the timely replacement of gut lining cells, skin surface cells, and blood cells, respectively. In contrast, other organs such as the liver show limited cell proliferation and very slow turnover. These organs are capable of cell division during the repair process following injury. Other cell types such as heart muscle cells (cardiomyocytes) and neurons are incapable of cell division during adulthood. Other cell types in these organs such as the connective tissue components of the heart and glial cells of the nervous system undergo cell proliferation and respond to injury and pathologic changes. The regulation of cell proliferation is therefore essential to understanding normal tissue, organ, and system function as well as the abnormal growth which occurs in tumors and other pathologic changes. Before discussing the regulation of the cell cycle, it is important to review the specific series of steps through which a dividing cell passes as it traverses the cell cycle.

PHASES OF THE CELL CYCLE

The cell cycle is the extremely precise mechanism by which two identical daughter cells are produced from a single parent cell with accurate duplication of the genetic material (DNA) and cytoplasm (organelles).

Cell populations which undergo rapid turnover such as an onion root or the small intestinal epithelium contain numerous mitotic figures. If these cells are observed with the light microscope, morphologic changes occurring in the nucleus and cytoplasm of these cells would be seen.

Although the cell cycle occurs as one process, different terms may be used to describe the processes of nuclear (mitosis) and cytoplasmic division (cytokinesis).

The genetic material in the nucleus must be duplicated, the nuclear envelope must breakdown for the duplicated chromosomes to come in contact with the machinery required for chromosomal segregation (spindle apparatus), the chromosomes must be separated to opposite poles of the spindle, cytoplasm must be shared between the two newly formed cells, and ultimately the nuclear envelope must reform in the two new daughter cells. These changes are associated specifically with individual phases of the cell cycle as outlined in the table and Fig. 10-1.

PHASES OF THE CELL CYCLE

PHASE OF CELL CYCLE	DEFINING EVENT(S)
Interphase (G_1, S, and G_2 phases)	Duplication of centrioles and DNA synthesis (S phase)
Prophase	Nucleolus disappears
Prometaphase	Nuclear envelope breaks down
Metaphase	Alignment of chromosomes in metaphase plate
Anaphase	Separation of sister chromatids; initiation of cytokinesis
Telophase	Nuclear envelope reforms; completion of cytokinesis

Interphase consists of the DNA synthetic phase (S) which is preceded by a gap period (G_1) required for the synthesis of cellular materials required for S phase and a gap period (G_2) preceding M, in which the materials required for mitosis are assembled.

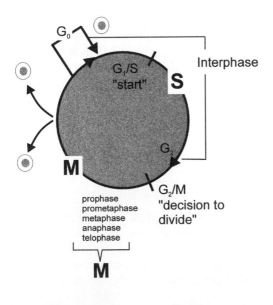

Figure 10-1 The cell cycle. Interphase consists of G_1, S, and G_2. M is divided into five substages: prophase, prometaphase, metaphase, anaphase, and telophase. Cells may leave the cell cycle and enter G_0, where they do not undergo RNA synthesis. Two critical control periods are illustrated: "start" (G_1/S interface) and the decision to divide (G_2/M interface).

DNA synthesis, the accurate duplication of the genetic material for eventual distribution to the two daughter cells, is the critical event in interphase, although other events such as duplication of the centrosome occur during this period. There are also two critical control points which occur within or abutting interphase. These control points are the G_1/S interface and the G_2/M interface, which control entry into S and M, respectively.

M phase consists of the five stages: prophase, prometaphase, metaphase, anaphase, and telophase.

The M phase, or mitotic phase, encompasses the events associated with the distribution of the duplicated DNA to the two daughter cells: disappearance of nucleoli, breakdown of the barrier between the cytoplasm and the chromosomes, assembly of the mitotic spindle, placement of the chromosomes on the spindle, separation of the chromosomes between the two daughter cells, and finally the cleavage of the cytoplasm between the two daughter cells.

G_0 is an extended "resting" phase defined by the absence of RNA synthesis.

Cells in G_0 may be recruited into the cell cycle on appropriate stimulation. For example, nondividing skin surface cells (keratinocytes) may be recruited into the cell cycle from the G_0 phase. The ability to recruit cells from G_0 is critical in repair and injury situations as well as in the response of tumors to radiation treatment and chemotherapy. In a tumor the radiation and chemotherapy primarily kill the dividing cells; the ability of the tumor to recruit nondividing cells into the cell cycle is directly related to its resistance to treatment.

The cell cycle can be separated into a series of cycles representing events required for cell division: centrosomal cycle, nuclear cycle, cytoplasmic cycle, and chromosomal cycle.

CENTROSOMAL CYCLE

The *centrosome* is a cytoplasmic structure consisting of matrix and, *in interphase, a pair of centrioles* oriented perpendicularly to each other and constructed primarily from microtubules (Fig. 10-2).

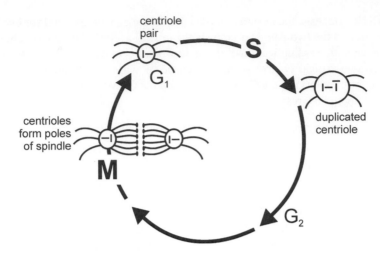

Figure 10-2 The centrosomal cycle.
Centrioles, within the centrosome, duplicate during interphase, form the poles of the spindles during mitosis and separate during telophase and cytokinesis.

The centrosome plays two essential roles in cell cycle events: (1) it contains a pair of centrioles which duplicate and form the poles of the mitotic spindle and (2) the centrosomes also nucleate microtubules to form the aster (radial array) of the spindle complex.

NUCLEAR CYCLE

The nuclear envelope represents a barrier between the chromosomes, which will be segregated into two newly forming daughter cells, and the cytoplasm, in which the mitotic assembly apparatus is constructed. It must break down to allow contact between the chromosomes to be moved and the microtubular system used to segregate the chromosomes to the two daughter cells. The nuclear envelope must subsequently be reformed to reestablish the segregation of the chromosomes within the nuclei of the newly formed daughter cells.

The nuclear envelope completes a cycle of breakdown and reformation orchestrated by the intermediate filament proteins called lamins.

Lamins are phosphorylated early in the cell cycle *leading to instability and dissolution of the nuclear envelope*; later, *dephosphorylation leads to reestablishment of the nuclear envelope.*

CYTOPLASMIC CYCLE

The cytoplasmic cycle involves the increase in volume and attainment of a critical mass (number of organelles) which must be reached for cell division to be completed.

If cytoplasmic volume is not sufficient to form two cells, then G_1 is delayed until that size is attained.

This phenomenon is critical to formation of identical cells within a tissue or organ. In some systems, however, the cytoplasm is reduced in each cell division by shortening of the G_1 period, allowing for no growth between successive cell cycles. This occurs in the cleavage divisions of the embryo following fertilization with the blastomeres resulting from cell division being many times smaller than the fertilized egg (zygote).

At a molecular level, the cytoplasmic cycle follows the rise and fall in concentration of a cytoplasmic oscillator called cyclin which regulates entry into the cell cycle.

The activity and regulation of *cyclin* is critical to cell cycle progression. As discussed later in the chapter, cyclins are combined with cell division cycle kinases (Cdks) to regulate entry into the cell cycle (passage from $G_1 \to S$) and entry into mitosis (passage from $G_2 \to M$).

CHROMOSOMAL CYCLE

The chromosomes are duplicated during the cell cycle, sister chromatids are separated into daughter cells, and the process begins again with a subsequent cell cycle.

The chromosomal cycle is the process involved directly in the duplication and distribution of the genetic material to two daughter cells. The process is carried out by DNA polymerases.

Chromosomes in mammalian cells are duplicated in a complex series of events including enzymes which open up the double helix (helicases), and two different polymerases which are associated with the leading and lagging strand templates.

MECHANICS OF CELL DIVISION: SPINDLE APPARATUS

A useful analogy is as follows: The cell cycle can be considered analogous to a cemetery burial in which the deceased is inactive and the center of events, with the chromosomes playing the role of the deceased. The pallbearers are analogous to the cellular machinery for separating the chromosomes and dividing the cytoplasm into two equal aliquots.

The active participants in mitosis are microtubules and the contractile ring. Microtubules form the spindle apparatus and are responsible for the separation of the chromosomes (anaphase), whereas the contractile ring of actin and myosin is responsible for the separation of the daughter cells (cytokinesis).

Microtubule growth during the formation of the spindle apparatus is an ideal example of microtubule assembly.

The mitotic spindle is formed by the addition of tubulin monomers to the microtubule organizing center (MTOC), which is the centrosome of the cell.

As discussed earlier, the *centrosome duplicates during interphase and each centrosome of the duplicated pair nucleates an aster* (a radial array) of microtubules which forms the two ends of the spindle. The dense matrix around the centrioles within the centrosome serves as the actual MTOC. The addition of tubulin (polymerization) or removal of tubulin (depolymerization) occurs through the process of treadmilling, which is discussed in Chap. 3, "Cytoskeleton."

The microtubules radiating from the centrosome are arranged with their minus ends closest to the centrosome and their plus ends toward the periphery (Fig. 10-3).

The microtubules associated with the spindle are of three types: (1) astral microtubules which radiate toward the cell periphery; (2) polar microtubules, which extend from pole to pole or aster to aster; and (3) kinetochore microtubules, which extend from one pole and attach to the kinetochore (a multiprotein complex) in the centromere region of the chromosomes. Kinetochore microtubules therefore attach chromosomes to the spindle apparatus (Fig. 10-4).

In metaphase, chromosomes are located in one plane, the metaphase plate. The chromosomes vibrate and move between the poles (asters) during prometaphase and reach an equilibrium position between the poles in metaphase. *The chromosomes are held in position by the kinetochore microtubules attached to opposite poles of the spindle apparatus.*

Anaphase is divided into two stages: (1) anaphase A, in which kinetochore microtubules shorten as depolymerization of tubulin occurs primarily at the kinetochore with the result that the chromosomes are pulled toward the poles, and (2) anaphase B, in which the polar microtubules elongate by tubulin polymerization and the poles of the spindle apparatus move further apart.

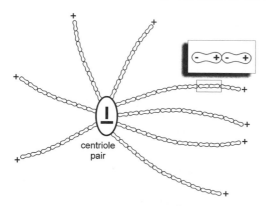

centriole
pair

Figure 10-3 The centrosome functioning as a microtubule organizing center (MTOC). Polarity is indicated by the plus (+) and minus (−) signs.

Kinetochore Microtubules

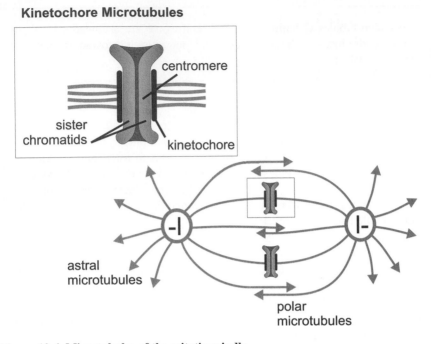

Figure 10-4 Microtubules of the mitotic spindle.
The three types of microtubules: astral, polar, and kinetochore.

Anaphase appears to be triggered by a specific signal in the form of an anaphase promotion complex (APC) which is discussed later in this chapter. Calcium is instrumental in the segregation of chromosomes as supported by the presence of calcium-containing vesicles immediately before anaphase and the precocious induction of anaphase by microinjection of calcium into a metaphase cell.

MECHANICS OF CELL DIVISION: CYTOKINESIS

In contrast to chromosome segregation, which occurs through the action of tubulin polymerization and depolymerization of microtubules, the separation of the two daughter cells (cytokinesis) occurs through the action of actin and myosin in the contractile ring (Fig. 10-5).

The contractile ring is active during anaphase and particularly in telophase and is responsible for the formation of the cleavage furrow which delineates the line of separation between the two daughter cells. This process of *cytokinesis* is

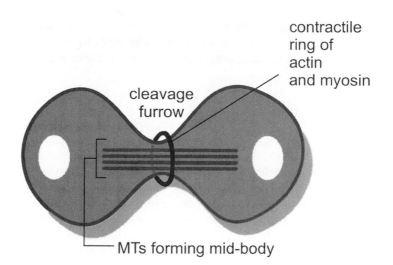

Figure 10-5 Cytokinesis.
Cytokinesis and the contractile ring. MTs = microtubules.

independent of microtubule function, but can be blocked by antimyosin or anti-actin antibodies, indicating the importance of actin-myosin interactions in the separation of the cytoplasm of the two newly formed cells.

REGULATION OF THE CELL CYCLE

The decision to enter the cell cycle is based on both internal and external cues. The entry point is known as *start* and is dependent on the cells possessing sufficient size and appropriate organelles. External cues include growth factors such as epidermal growth factor (EGF), platelet-derived growth factor (PDGF), and fibroblast growth factor (FGF), which enhance cell proliferation, and transforming growth factor beta (TGFβ), which inhibits cell proliferation in most cell types.

There are three checkpoints in the cell cycle: G_1/S (start), G_2/M (the decision to enter mitosis), and regulation within M phase (e.g., alignment of metaphase chromosomes and anaphase segregation of chromosomes).

The two best studied control points are regulation of start and mitosis. The mitotic checkpoint has been studied most extensively. Initial experiments using synchronized cells in different phases of the cell cycle demonstrated that mitotic cells contained a factor capable of driving a cell in any other stage of the cell cycle into mitosis. This factor was later found to be *MPF, the mitotic promoting factor*, which drives the phosphorylation of proteins associated with mitosis (Fig. 10-6).

MPF consists of two parts: cyclin and a cyclin-dependent kinase (Cdk).

The kinase is the portion of the molecule which is phosphorylated during M phase. This is the high phosphorylation state or upstroke in the carousel and pony analogy. Interphase is the low phosphorylation state. *Cyclin is a cytoplasmic oscillator which increases in concentration during the interphase stage of the cell cycle.* Cyclin is degraded by *ubiquitination* late in the mitotic phase, which causes dissolution of the MPF (Fig. 10-6). *Ubiquitin* is attached to a "destruction box," or destruction sequence, on cyclin. Addition of multiple ubiquitins leads to formation of a *proteosome* containing the degradative proteases. The relationship of the cyclin cycle to G_2/M control is shown in Fig. 10-7.

A similar regulatory complex of G_1 cyclin plus Cdk (cell division kinase) controls entry into the S phase.

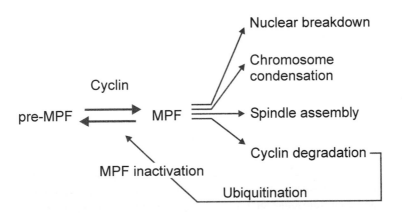

Figure 10-6 MPF-induced phosphorylation.
MPF drives the phosphorylation of proteins associated with mitosis. Ubiquitination is the process used for cyclin degradation.

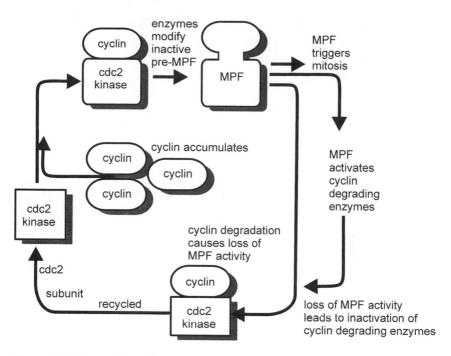

Figure 10-7 The cyclin cycle.

Regulation of the start point follows the same principles of Cdk phosphory-lation and cyclin degradation during the end of S phase and is similar in regulatory pattern to the M phase regulation. Growth factors trigger cascades of intracellular signals leading to changes in the cyclins, Cdks, and MPF.

> The activation of the cyclin-Cdk complex is regulated by a series of kinases and phosphatases (Fig. 10-8).

For example, the mitotic cyclin contains both an inhibitory and an activation site. Phosphorylation of the activation site occurs through the action of the activation kinase (MO15) while phosphorylation of the inhibition site occurs through the action of the inhibitory kinase (Wee1). The complex remains inactive until a specific phosphatase (cdc25) removes the phosphate from the inhibition site activating MPF.

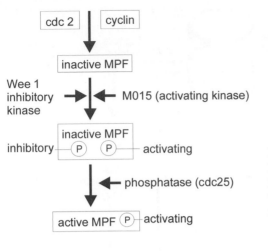

Figure 10-8
Regulatory phosphatases and kinases forming a cascade controlling activation of MPF. Cdc is an abbreviation for cell-division-cycle genes.

REGULATION IN M PHASE

Very little is known about this checkpoint of the cell cycle, although it has been shown that correct alignment of the chromosomes is required for normal cell cycle progression. Anaphase-promoting complex (APC) is required for progression through anaphase. APC is an enzyme with proteolytic activity which appears to act on a noncyclin substrate which normally inhibits anaphase events (segregation of chromosomes to separate cells).

CELL CYCLE EXPERIMENTS

Cell fusion experiments have been instrumental in providing information on the regulation of the cell cycle. In these classic experiments, cells in different stages of the cell cycle are fused using a viral mechanism. The experiment allows for the determination of the dominance of one cell cycle phase over another. These studies demonstrate which phases of the cell cycle contain regulatory factors for the control of cell cycle events.

Example 1

Cell in G_1 fused with a cell in S → G_1 cell is driven into S, and the S-phase cell continues through DNA synthesis.

Interpretation: The S-phase cell contains an activator which precociously drives interphase cells into DNA synthesis. The G_1 cell contains no factors which influence the S-phase cell.

Cells in S-phase contain a cyclin plus a Cdk in a complex. The presence of this complex in the S-phase cell directs the G_1 cell prematurely into DNA synthesis. The S-phase cell with incompletely replicated DNA is not regulated by the G_1-phase cell.

Example 2

Cell in G_1 fused with a cell in $G_2 \rightarrow G_1$ cell follows its normal timetable, whereas the G_2 cell remains in G_2 until the G_1 cell completes S and enters G_2 also.

Interpretation: The G_2 cell no longer contains the S-phase activator, or it has been inactivated. The G_1 cell cannot rereplicate its DNA.

The complex of cyclin and Cdk formed in S phase is degraded by the breakdown of cyclin. The cell-division-cycle (cdc) kinase is reutilized later in the next cell cycle. The degradation process is complete by the time the cell enters G_2. The G_2-phase cell is incapable of reentering the S phase because of a mechanism known as the rereplication block. The DNA replication process is carried out by replication forks which modify the segment of chromatin they have replicated. This modification of the chromatin constitutes the rereplication block. The rereplication block is removed after M phase, again allowing for the duplication of the DNA.

Example 3

A cell in G_2 fused with a cell in $S \rightarrow G_2$ cell will enter S, and the two cells will go through the cell cycle in synchrony.

Interpretation: The S-phase cell cannot enter G_2 until its DNA replication is complete.

Incompletely replicated DNA prevents entry of a cell into mitosis. A similar situation occurs after radiation damage which stimulates DNA repair and prevents mitosis. The p53 protein, which mediates damage-mediated inhibition of mitosis, is discussed later in the chapter.

Example 4

An interphase cell (G_1, S, or G_2), is fused with a M-phase cell → M-phase cell induces the interphase cell to enter mitosis.

Interpretation: The M-phase cell contains a dominant factor which overrides all other regulatory factors.

This factor is mitotic promoting factor (MPF) and is present in the cytoplasm of cells undergoing mitosis or meiosis. MPF is a potent phosphorylation factor and drives mitosis by phosphorylation of numerous proteins. Some of the key proteins are the lamins. Phosphorylation of the lamins, like that of other intermediate filament proteins, leads to destabilization. Destabilization of the lamins causes the breakdown of the nuclear envelope which occurs during prometaphase of the cell cycle (see previous table). Phosphorylation of the spindle proteins is also driven by MPF as is the phosphorylation of the histones leading to condensation of chromatin (Fig. 10-6).

REGULATORY FACTORS

Tumor Suppressors

Tumor suppressors function through transcription factors or through sequence-specific inhibitory regions. Removal of the suppressor leads to cell proliferation and release from the nonproliferating, or G_0, state.

Tumor suppressors were discovered through their effects on cancerous or transformed cells which no longer obey normal cell rules of proliferation and activity. For example, normal cells stop dividing *in vitro* (in culture) when their cell membranes all touch (a confluent culture). Transformed cells continue to proliferate and overgrow one another, ignoring the principle of *contact inhibition* exhibited by normal cells.

Retinoblastoma (Rb) gene product is an inhibitory factor which in the "resting" G_0 phase binds to a member of the E2F transcription factor family. When phosphorylated, the Rb gene product "permits" cell proliferation (Fig. 10-9).

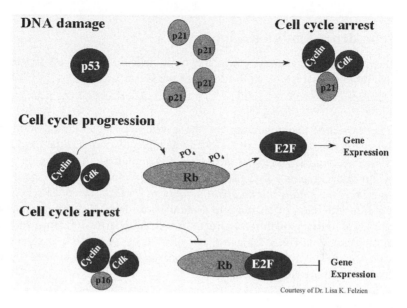

Figure 10-9 Cell cycle regulation by p53, retinoblastoma gene product, and related factors.
p21 is a family of inhibitors which interacts with Cdk-cyclin complexes; p16 is another family of inhibitors which regulate the G_1/S interface. E2F is the transcription factor family which regulates the cell proliferation genes. (*From Dr. Lisa K. Felzien.*)

Some cell proliferation genes contain an E2F transcription factor region which is inhibited by the unphosphorylated Rb gene product. Phosphorylation of Rb gene product results in the release of the binding and the activation of the transcription factors associated with cell proliferation genes. The result is elimination of the inhibitory influence of Rb and transcription of cell proliferation genes (Fig. 10-9).

p53 is a sequence-specific inhibitory gene regulatory product which inhibits the transcription of cell proliferation genes; it is a mediator of DNA damage.

p53 is activated by DNA damage which results in the delay in mitosis which occurs in a cell with damaged DNA. When DNA repair is complete, p53 activity decreases and mitosis can occur (Fig. 10-9).

Cyclin-dependent kinase inhibitors

Other important regulators of the cell cycle include the CdkIs, cell division cycle inhibitors which block the action of the Cdks.

The existence and regulation of the cyclin-dependent kinase inhibitors (CdkIs) increases the complexity of cell cycle regulation. CdkIs have been separated into families based on their generally preferred site of action [G_1/S (G_1-cyclin-Cdk complex) or G_2/M (mitotic cyclin-Cdk complex)]. These *proteins bind to and inhibit the catalytic activity of cyclin-Cdk complexes*. They arrest cell proliferation in response to a number of inhibitory stimuli including anti-cell proliferative growth factors (e.g., transforming growth factor beta), differentiation cues, and DNA damage. The CdkIs also play a role in regulation of normal cell cycle progression through the timing of their inhibition of the cyclin-Cdk catalytic activity (Fig. 10-10).

Growth Factors

Some growth factors stimulate cell proliferation (fibroblast growth factor, platelet-derived growth factor, and epidermal growth factor EGF), whereas other growth factors are primarily inhibitory in terms of their proliferative effect (TGF). These factors bind at the cell surface to specific receptors and induce a response through one or more signal transduction pathways (discussed in Chap. 9), "Cell Receptors and Intracellular Signaling".

In the case of EGF, binding occurs to a transmembrane cell surface receptor (EGF-R) on its target cell. In epidermis, EGF will bind to EGF-R on a keratinocyte (epidermal cell). EGF acts via a tyrosine-specific protein kinase pathway to trigger a cascade of phosphorylation. Involved in the signal transduction are cytoplasmic proteins such as Ras which transduce the tyrosine kinase activity into a nuclear signal. Cell proliferation is regulated intracellularly by mole-

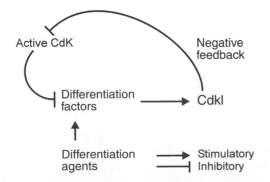

Figure 10-10 Role of cell cycle–dependent kinase inhibitors (CdkIs) in regulation of cell cycle events. CdkIs act through negative feedback to decrease the activity of cell cycle–dependent kinases (Cdks).

cules called *protooncogenes* because they are the normal cellular counterparts of genes associated with the uncontrolled cell proliferation of malignancy and tumor cell transformation (*oncogenes*). One of these protooncogenes is *myc*. This is an early response gene which is turned on soon after growth factor exposure. The cyclins and cdc kinases (or Cdks) which specifically control the cell cycle are delayed-type response genes far downstream from the initial phosphorylation cascade (Fig. 10-11).

Meiosis

Like the Krebs cycle in biochemistry, meiosis is a topic that is often taught, but quickly forgotten by most students. The problem is that meiosis is more complex than mitosis, and the process of meiotic cell divisions during the formation of the gametes within the testes and the ovaries is more difficult to visualize than the cell division cycle. Scientists also have a less clear understanding of meiotic events than the stages of mitosis.

Meiosis is essential to sexual reproduction and procreation. Meiosis takes diploid cells with homologous pairs of chromosomes (one from daddy and one from mommy) and allows for recombination between genetic information provided by mom and dad for each chromosome. This first part of meiosis is critical to the development of diversity. Subsequently, the gametes (sperm and eggs) develop as a result of meiosis. This second part of meiosis results in a halving of the ploidy number from the usual diploid state by providing only one chromosome of each homologous pair to the haploid cell. In the case of autosomes (the 22 non-sex chromosomes for humans), each haploid cell normally contains one

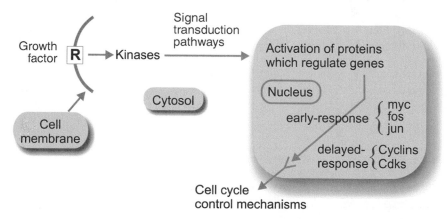

Figure 10-11 Growth factor regulation of cell proliferation.
Binding of a growth factor to its receptor initiates a series of cytosolic kinases and signal transduction pathways. The early-response genes (myc, fos, and jun) are turned on first, followed by the delayed-response genes such as the cyclins and Cdks (or cdc kinases).

homologue, either paternal or maternal. For the sex chromosomes meiosis results in each sperm being either an X or a Y sperm and the eggs being all X. This arrangement allows for the reconstitution of the diploid state at fertilization when an X or a Y sperm fertilizes an egg containing an X chromosome resulting in the development of a male or female zygote.

Meiosis is divided into two parts: the first meiotic division (I), in which crossing over and exchange of maternal and paternal information occurs between sister chromatids, and the second meiotic division (II), which is essentially a mitotic cell division cycle to form the haploid gametes.

Meiosis I is much more complex than meiosis II. Meiosis I requires the accurate recognition and pairing of maternal and paternal chromosomes to form a complex structure for the exchange of genetic information. Meiosis II establishes a metaphase plate similar to mitosis in which the homologous pairs separate into individual gametes. Meiosis I and II are discussed separately here.

Meiosis I is divided into stages which are similar to mitotic stages: interphase, prophase, metaphase, and anaphase.

Interphase consists of G_1, S, and G_2 as in mitotic interphase. During S phase duplication of DNA occurs in the parental (diploid) cell so that each chromosome is represented by two pairs of sister chromatids. There is a segment of DNA which remains unreplicated during S; this segment, known as zygDNA, may function as a vice-grip to hold the two chromatids of each chromosome in close proximity during meiotic prophase. The next stages of meiosis represent the preparations (alignment of the chromosomes) for genetic recombination: the exchange (crossing over) of bits of genetic information between nonsister (maternal and paternal) chromatids (Fig. 10-12).

Prophase I is the critical stage in meiosis, is very complex, and consists of five substages: leptotene, zygotene, pachytene, diplotene, and diakinesis.

It is during prophase I that the chromosomes pair and recombination of genetic material occurs. The substages of prophase represent the transition from

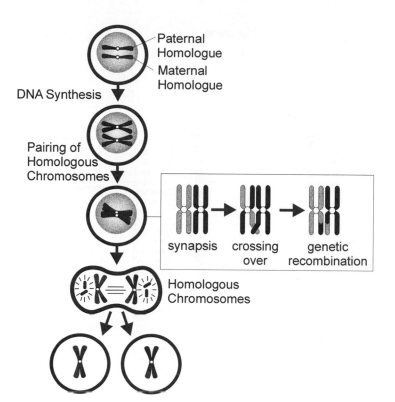

Figure 10-12 Meiosis I.
The steps involved in the first stage of meiosis. The key event is recombination, which
provides the genetic variation that occurs in the germ cells.

separate maternal and paternal sister chromatids to the assembly of the *synaptonemal complex, which facilitates the exchange of genetic information between
maternal and paternal chromatids.*

> The synaptonemal complex is the centerpiece of prophase I like the spin-
> dle apparatus which is the critical structure in the mitotic cell division
> cycle. The synaptonemal complex provides a means of splicing together
> the crossed over strands from maternal and paternal chromatids
> (Fig. 10-13).

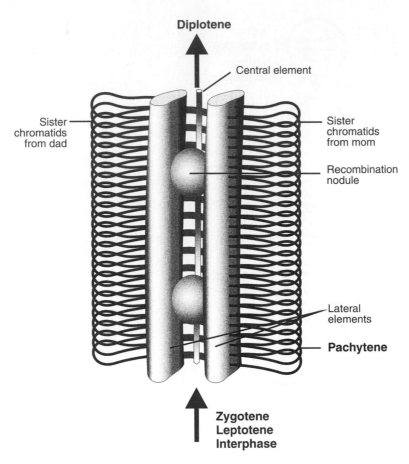

Figure 10-13 The synaptonemal complex.

The key events of each substage phase of the first meiotic division are listed in the table. The stages can be remembered with the famous mnemonic: LoZangelesPoliceDepartmentDummy (L = leptotene, Z = zygotene, P = pachytene, D = diplotene, and D = diakinesis).

The stages of meiosis all center around the formation and breakdown of the synaptonemal complex. *Synapsis is the process of pairing which finally results in the alignment of the two chromosomes of each homologous pair along their entire length.*

KEY EVENTS OF EACH SUBSTAGE
OF THE FIRST MEIOTIC DIVISION

SUBSTAGE OF PROPHASE I	EVENT OCCURRING IN SUBSTAGE
Leptotene	Initial condensation of the chromosomes, initiation of assembly of the central component of the synaptonemal complex.
Zygotene	Continued formation of the synaptonemal complex; chromosomes pair.
Pachytene	Synapsis complete; recombination occurs.
Diplotene	Beginning of disassembly of the synaptonemal complex (initiation of desynapsis): transcription of RNA.
Diakinesis	Completion of disassembly (desynapsis); condensation is complete.

Pachytene is the stage which assures that each person is not a clone, by facilitating the process of homologous recombination through synapsis.

The structure of the synaptonemal complex facilitates recombination, but the process of recombination requires the presence of recombination nodules (large multiprotein complexes) which move along the homologous maternal and paternal chromosomes to facilitate chromatid exchanges. The structure of the complex is analogous to ping pong balls rolling along a ladder supported by two straws (lateral protein segments) with rungs formed by toothpicks. The ping pong balls are protein complexes which facilitate chromatid exchange by an unknown mechanism. This exchange process occurs in pachytene.

Complementary-DNA base pairing is believed to be involved in the recognition of homologous chromosomes.

Metaphase I involves the separation of paternal and maternal chromosomes to individual cells, and the mechanism differs from mitotic metaphase.

The structural arrangement of meiotic metaphase I differs from mitotic metaphase in that kinetochores of sister chromatids fuse and the chromatids are also fused at the chiasmata (points of physical contact between duplicated homologous chromosomes).

In anaphase I the random sorting of maternal and paternal homologues between two diploid cells is complete.

Kinetochore fibers pull the paternal or maternal chromatids into separate cells. The two diploid cells will subsequently undergo a mitotic division (meiosis II) to form the gametes with half the DNA and only one sister chromatid of the maternal or paternal pair.

Meiosis II consists of interphase II, prophase II, metaphase II, anaphase II, and telophase II.

The events of meiosis II are shown in Fig. 10-14. These stages closely resemble the stages of mitosis except for the fact that *interphase is very short because of the absence of DNA synthesis.*

The details of meiosis as it pertains to specific cell types in the ovary and testis are discussed in Chap. 25.

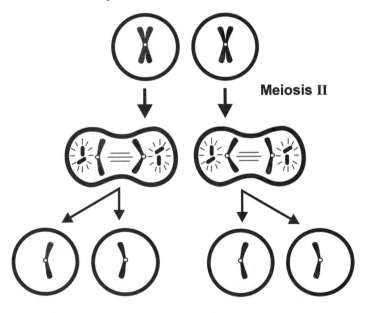

Figure 10-14 Meiosis II.
The steps involved in the second stage of meiosis. This second meiotic division reduces the ploidy number from 2n (diploid) to n (haploid), creating the haploid gametes. The diploid number is restored if fertilization occurs.

BASIC HISTOLOGY

·

· · · · · · · · · · · ·

Histology is the study of the cells, tissues, and organs of the body using both the light and electron microscope. Following the first part of this text, which reviews fundamental concepts of cell biology, this second part of the book presents the basic concepts of histology, the study of cell, tissue, and organ structure and function.

The cell is the fundamental unit of the body. Part II reviews the organization of isolated cells as well as cells coupled with other cells to form tissues. The basic tissues of the body are epithelium, connective tissue, muscle, and nervous (neural) tissue. The basic concepts of structure and function are presented for each of these tissues. Organs contain tissues of different types organized into different arrangements. The organs work together as functional systems to carry out specific roles for the organism as a whole. The functions carried out by these systems include protection of the body from foreign invaders (immune and integumentary systems), obtaining oxygen and removing carbon dioxide (respiratory system), transport of the blood around the body (circulatory system), digestion (gastrointestinal system and related glands), formation and removal of liquid waste in the form of urine (urinary system), integration of body systems with each other and the environment (nervous and endocrine systems, as well as the special senses), and procreation (reproductive systems). Organs are organized into functional systems (immune, circulatory, gastrointestinal, urinary, endocrine, reproductive, nervous) to serve major functions for the organism as a whole. The basic concepts of each of these systems in the body hierarchy are reviewed, followed by a critical section to assist the student in developing the fundamental diagnostic skills required for tackling the routine identification of tissues and organs under the light microscope. The last chapter deals with the basic tools of the histologist or cell biologist. That chapter may seem like an appendix, but is essential to the overall concepts of size and resolution, and for understanding the structural and functional relationships of cell and tissue biology.

· · · · · · · · · · · ·

CELLS ARE US

·

· · · · · · · · · · · ·

This chapter provides the basic histologic concepts of the cell at the light and electron microscopic (ultrastructural) levels. In Chap. 28, "Strategies for Tissue and Organ Identification," the key features to be used in microscopic recognition of histologic tissues and organs are highlighted in an easy to follow flowchart method. In Chap. 29 an overview of the tools of histology is presented. The fundamentals of the light and electron microscopes and the basics of staining and a few critical cell biologic methods are provided in that chapter as well.

ZEN OF THE CELL

Why does a particular cell have extensive rough endoplasmic reticulum? Why is there an abundance of mitochondria in the cytoplasm of other cells? Why is there extensive smooth endoplasmic reticulum in some cells? Cell structure reflects the specific function carried out by a cell. This is the emphasis of the authors and should be the emphasis of students. If structure and function are treated as two separate and unrelated parts, it will be more difficult to learn each. The most important aspect of histology and cell biology is the relationship between cell structure and function.

The best way to develop the basic concepts of the cell from a histological perspective is to provide a diagram as an overview of what can be seen in an electron micrograph. There is no single cell which shows all organelles and cell structures optimally. The closest example of a generalized cell is the liver cell (hepatocyte). A diagram of an electron micrograph of a hepatocyte is shown on the next page (Fig. 11-1).

While looking at the electron micrograph, remember that tissue for electron microscopy is stained with heavy metals such as lead and osmium. This method enhances the differences in electron density and electron lucency as an electron beam passes through the tissue in the transmission electron microscope.

Starting with the outside of the cell, the *membrane* is visible only because of the osmium staining. Why does the membrane appear as a trilaminar structure (two electron-dense layers sandwiching an electron-lucent layer)? The dense layers represent the hydrophilic heads of the phospholipid bilayer with the hydrophobic (lipophilic) tails forming the contents of the electron-lucent sandwich. The bread of the sandwich is covered with a sandwich spread of fatty acid tails and some integral and peripheral proteins (so much for the low-fat spreads). At least the cholesterol is left for the bread layer.

Next, look for the *nucleus*. If the nucleus is included in the electron micrograph, it will be seen as an ovoid structure bounded by a double nuclear envelope. Within the nucleus, a dense granular structure known as the nucleolus may be visible. Chromatin may also be visible in its two forms: (1) dense granular clumps along the periphery of the nuclear envelope (inactive chromatin, heterochromatin) and (2) the granular electron-lucent chromatin found throughout the nucleus (active chromatin undergoing transcription, euchromatin).

Then look for the *mitochondria*. These organelles are surrounded by a double membrane and vary considerably in number and shape, depending on the cell type. Typically, the mitochondria have a cigar shape with an electron dense sandwich-membrane arrangement with finger-like extensions of the inner mitochondrial wall called *cristae*.

Throughout the cytoplasm look for flattened tubules, cisternae, and saccules which make up the *endoplasmic reticulum*. The endoplasmic reticulum may be studded with ribosomes, electron-dense dots along the membrane (rough endoplasmic reticulum, RER). The RER has two surfaces or compartments, the cisternal and the cytosolic. Proteins destined for export must be translocated into the cisternal space (lumen) of the RER. The smooth endoplasmic reticulum (SER) is devoid of ribosomes. If you see extensive SER what does it mean? The SER is involved in steroid synthesis and detoxification, so steroid-producing endocrine cells and detoxifying liver cells contain a high concentration of SER.

There are smaller and more distended saccules in the shape of a half-moon which form the *Golgi apparatus*. The saccules usually number three to five. Why would a cell have extensive amounts of Golgi? The Golgi apparatus is involved in packaging of protein for secretion and adds sugars to molecules which contain both sugars and proteins.

Around the periphery of the cell look for *vesicles*. These are usually of varying density and size. Secretory vesicles tend to increase in density as one moves toward the surface at which their contents will be released.

Still smaller are the *granular arrays of glycogen* appearing as irregular black dots. They provide a light stippled background for many electron micrographs.

Lipid droplets also vary in size and electron density, but they tend to be lighter staining than most secretory vesicles and lack a membrane boundary around their surfaces.

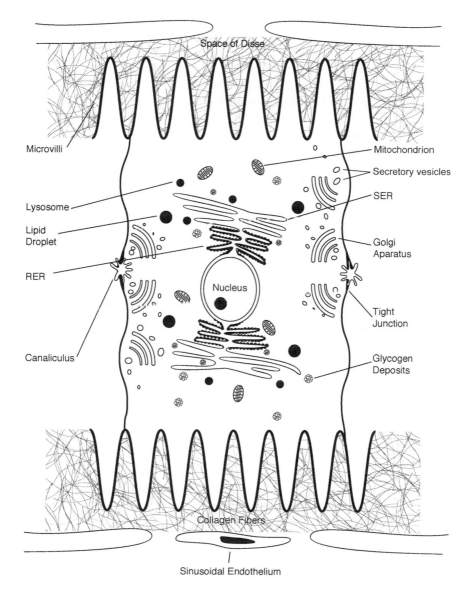

Figure 11-1 Subcellular structures.

There are three types of tubules or filaments in the cell: microtubules, intermediate filaments, and microfilaments, in order of decreasing size (diameter). These structures may be cut longitudinally or in cross-section.

Lysosomes are another membrane-bound organelle with an amorphous granular material filling the center of the organelle. Electron-dense material inside the lysosome usually represents a work in progress, partially digested organelles. One size factor to keep in mind is that lysosomes are generally smaller than mitochondria and more spherical. Also, there are no cristae in the lysosomes.

Peroxisomes are abundant in some cells such as hepatocytes which play a major role in breakdown of long-chain fatty acids and ethanol. The peroxisomes are similar in size to lysosomes, but contain a fine, dense, granular content.

It is important to note that some subcellular structures are visible at the light-microscopic level. For example, the nucleus, nucleolus, and chromatin can be seen at the light-microscopic level. The mitochondrion at 0.5 to 1.0 μm in size is at the boundary of the resolution level of the light microscope. Mitochondria may be seen as dots when tissues are stained with special stains. Although other structures cannot be seen individually, some organelles can have sufficient concentration to be detected under the light microscope: rough endoplasmic reticulum in the pancreatic acinar cell and the motor neuron where it forms the Nissl substance. Secretory granules can also be seen in some cells because of the high density of vesicles in a particular part of the cell. However, it is important to appreciate size and resolution issues: individual secretory granules, ribosomes, or glycogen deposits cannot be visualized with the light microscope.

EPITHELIUM: LININGS UNLIMITED

·

· · · · · · · · · · · ·

Why do some organs have an epithelial covering? What does an epithelium do? Why are epithelia needed? Some organs are covered by capsules which provide a protective coating for the organ. Although epithelia also function in a protective mode, they carry out other specific functions related to the overall organ function. It is important to think of the epithelium in context of the organ function, one of those structure-function relationships that are part of the basic concepts emphasized throughout this text.

The epithelium covers the outside of organs or structures (the epidermis of the skin) or it lines various internal tubes of the body (gastrointestinal tract, ducts, or blood vessels). Even in these cases, the type of epithelium varies based on function of the structure. Epithelial functions include protection, secretion, absorption, stretch, diffusion, and active transport. Epithelial cells are not free agents and have little mobility except within the epithelium. The citizens of epithelia are tightly bound to each other and to their underlying substratum which they synthesize. Epithelia invariably undergo cell turnover with new cells being born and replacing older cells which are sloughed off at the surface.

JUNCTIONS BETWEEN EPITHELIAL CELLS

The epithelium has a minimum of intercellular substance and a close association between adjacent cells.

The components of the junctional complex have been discussed previously in Chap. 7, "Cell-Cell, Cell-Matrix Interactions." Simple columnar epithelia of the gastrointestinal tract exhibit all the classic components of the junctional complex: *zonula occludens* (tight junction), *zonula adherens* (intermediate junction), and the *macula adherens* (desmosome). At the light microscopic level, the entire junctional complex is visible as a thin dense line known as the *terminal bar*. This should not be confused with the terminal web, which is the cytoskeletal network visible as a dense area in the apical portions of the epithelial cells (see Chap. 3, "Cytoskeleton").

POLARITY OF EPITHELIAL CELLS

The epithelial cell has a distinct polarity with apical, basal, and lateral surfaces constituting distinct domains.

Polarity is a method for specialization of cell surfaces for specific functions. For example, receptors and other proteins need to be specifically targeted to the cell surfaces where they function. The junctional complexes of the lateral domains provide a morphologic separation of the apical and basolateral domains.

INTERRELATIONSHIP OF JUNCTIONAL COMPLEXES AND EPITHELIAL POLARITY

The structure and function of the junctional complexes and the polarity of epithelia are inexorably linked. First, the junctional complex is the unique component of the lateral domain. Second, the junctional complex, particularly the zonula occludens, maintains apical versus basolateral polarity.

The apical domain can be identified by labeling with specific lectins (sugar-binding molecules). With junctional complexes intact, lectin staining for the apical domain is specific. When the junctional complexes are disrupted and the cells separated, apical-specific lectin binding is lost and lectin staining becomes uniformly distributed around the cell (Fig. 12-1).

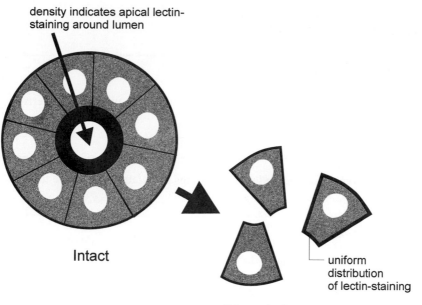

density indicates apical lectin-
staining around lumen

Intact

uniform
distribution
of lectin-staining

Disrupted

Figure 12-1
Junctional complexes maintain the apical-basolateral polarity of the cell. When junctional complexes are disrupted, apical markers (lectins) are found distributed uniformly around the separated cells.

LATERAL DOMAIN

The lateral domains of epithelial cells are distinctive and contain the junctional complexes. The lateral domain is often lumped with the basal domain to constitute the basolateral domain.

BASOLATERAL DOMAIN

Basolateral domains possess Na^+,K^+-ATPase activity associated with active transport of ions. This enzyme is responsible for the low intracellular concentration of sodium ions within the cell.

Na^+,K^+-ATPase enzymatic activity is Mg^{2+}-dependent and is the biochemical basis of the sodium pump which transports sodium from the cell, maintaining the required low Na^+ concentration of the intracellular environment. Water and Cl^- follow Na^+ out of the cell.

BASAL DOMAIN

The basal surface of the epithelial cell does have some unique characteristics.

The basal domain is modified with infoldings associated with active transport.

These basal infoldings, which are seen in cells like the tubular epithelial cells in the kidney or the striated ducts of the salivary glands, are involved extensively in the active transport of ions. The infoldings represent plasma membrane folds which isolate numerous mitochondria stacked in between the folds. At the light microscopic level the infoldings appear as striations. Hence the name *striated ducts* in the salivary glands.

LINKAGE TO THE UNDERLYING SUBSTRATUM

The basal surface of the epithelium is also associated with the basal lamina which it produces.

The basal lamina provides a scaffolding and is composed of a selective mortar which connects the epithelium with the underlying connective tissue.

The basal lamina contains a specialized type of collagen called *type IV* (see Chap. 13) and critical extracellular matrix molecules like *fibronectin* and *laminin*, which link to integrins (fibronectin and laminin receptors; see Chap. 7, "Cell-Cell, Cell-Matrix Interactions") across the basal membrane of the epithelial cell. The term *basal lamina* is used to describe the components of the underlying substratum, synthesized by the epithelium, which can be seen only with the electron microscope. Those structures which can be seen with the light microscope are usually referred to as the *basement membrane*. To be visible with the light microscope, a special stain which visualizes glycoprotein (periodic acid–Schiff, PAS) must be used. What the observer sees in such a preparation is either a double

basal lamina or the basal lamina plus the associated extracellular material produced by the underlying connective tissue cells.

APICAL DOMAIN

The apical domain of the epithelial cell has a number of important specializations: (1) cilia, (2) microvilli, and (3) stereocilia.

The apical domain of epithelial cells is modified for absorption in the case
of the microvilli and stereocilia (modified microvilli). Cilia represent an apical
specialization for movement of materials over the cell surface.

Microvilli are finger-like projections which dramatically increase the apical surface area for enhanced absorption.

At the ultrastructural level, the microvilli appear as fingers extending from
the cell, covered by a *glycocalyx* (cell coat) consisting of glycoproteins and glycolipids. Individual microvilli cannot be seen at the light microscopic level. The
glycocalyx plus microvilli stain as a lighter border (*brush border*) with standard
stains, but are seen distinctly with periodic acid–Schiff (PAS) stain, which stains
glycoproteins and glycolipids.

Stereocilia are long, immotile, modified microvilli found in organs such as
the epididymis where they function in absorption.

Stereocilia are a misnomer. Their name would imply that these structures are
involved in motility and have a similar structure and function to cilia. However,
stereocilia are immotile and have no structural similarity to cilia. They are
involved in absorption of testicular fluid in the epididymis. In the ear they function as sensors of cochlear hair cells in that tilt or deflection results in depolarization.

Cilia are motile and move fluid and materials over the cellular surface.

The structure of cilia is discussed in Chap. 8, "Cell Motility." Cilia possess
the classic "9 + 2" arrangement of paired microtubules. The cilia play a key role

in the function of the mucociliary escalator, part of the defense mechanism associated with the respiratory system (Chap. 20).

HISTOLOGY OF THE EPITHELIUM

Nomenclature

Epithelia are divided into two main types, simple (composed of a single
layer) and stratified (composed of two or more layers).

Simple epithelia are always found lining lumena of ducts, blood vessels, and
other tubelike structures of the body. The single layer of epithelium facilitates
transport and presents a minimal barrier while still maintaining epithelial polarity. Stratified epithelia serve to line lumena, but also provide enhanced protection
on outside surfaces such as the skin.

Pseudostratified epithelium is an in-between kind of epithelium. It appears
stratified because nuclei are found at different levels, but is really a single
layer of cells in which all cells touch the basal lamina, but not all cells
reach the free surface.

Pseudostratified epithelium is often described as consisting of basal cells
and columnar cells, all of which reach the basal lamina. The epithelium is ciliated
in the respiratory system and found without cilia in the male reproductive system.

Simple epithelia are named according to the shape of the cells: squamous
(flattened), cuboidal (cubes), and columnar (columns).

The shapes of simple epithelia and the arrangement of nuclei in each are
shown in Fig. 12-2.

Stratified epithelia are classified on the basis of the shape of the surface
layer. The stratified epithelia defined in this way are stratified squamous
(multiple layers), stratified cuboidal, and stratified columnar (the latter two
only two layers thick).

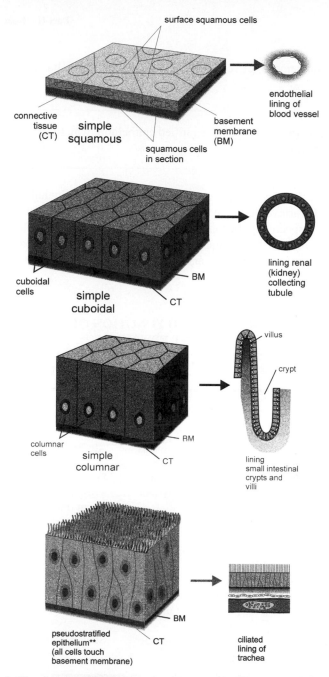

surface squamous cells

connective tissue (CT)

simple squamous

squamous cells in section

basement membrane (BM)

endothelial lining of blood vessel

cuboidal cells

simple cuboidal

BM

CT

lining renal (kidney) collecting tubule

columnar cells

simple columnar

BM

CT

villus

crypt

lining small intestinal crypts and villi

pseudostratified epithelium** (all cells touch basement membrane)

CT

BM

ciliated lining of trachea

Figure 12-2 The simple epithelia.

The left column shows the structure of the epithelial type and the right column a specific example (location) of the epithelial type. Note that the pseudostratified epithelium is classified as a simple type because it consists of a single layer of cells. All the pseudostratified epithelial cells touch the basement membrane, but all the cells do not reach the lumenal surface.

157

The arrangement of cells in the stratified epithelia are shown in Fig. 12-3.

There is one additional type of stratified epithelium, which is called transitional epithelium.

The *transitional epithelium* is found in the urinary tract including the ureters and bladder. It varies from three to six layers for the stretched (distended) to the relaxed state, respectively. In contrast to stratified squamous epithelium, the surface cells of transitional epithelium are large and rounded in the relaxed state.

The characteristics, location, and function of the different types of epithelia are summarized in the table below.

LOCATION AND FUNCTION OF
DIFFERENT EPITHELIA

EPITHELIAL TYPE	LOCATION	FUNCTION
Simple squamous	Lining blood vessels	Transport
Simple cuboidal	Renal collecting duct	Secretion, absorption, transport
Simple columnar	GI tract from stomach to colon	Absorption, secretion
Pseudostratified	Conducting portion of respiratory system	Protective, cilia function as a respiratory defense mechanism (mucociliary escalator, or tracheobronchial escalator)
Stratified squamous	Skin, esophagus, anus	Protective in areas of high friction
Stratified cuboidal	Sweat ducts	Transport
Stratified columnar	Salivary gland ducts, inner layer of the conjunctiva	Transport
Transitional	Urinary tract: bladder and ureters	Stretch, barrier

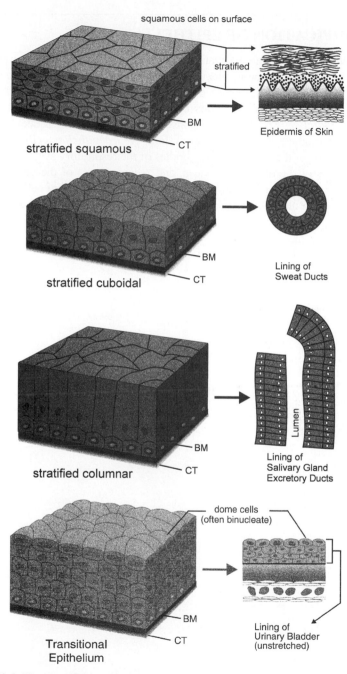

Figure 12-3 The stratified epithelia.
The left column shows the structure of the epithelial type and the right column a specific example (location) of the epithelial type.

MODIFICATION OF EPITHELIAL CELLS

Epithelial cells in different organ systems are modified to form (1) myoepithelial cells and (2) neuroendocrine (neuroepithelial) and enteroendocrine cells. The myoepithelial cells contain actin and myosin and wrap themselves around the secretory units of the salivary and mammary glands. *Myoepithelial cells* contract in response to secretagogues by squeezing the secretory product(s) into the duct system. *Neuroepithelial cells* include sensory cells of the neuroepithelial bodies of the respiratory systems, bipolar neurons of the olfactory mucosa, and receptor cells of the taste buds. The *enteroendocrine cells* are derived from an epithelial stem cell and differentiate into unicellular endocrine glands releasing a variety of peptides and other hormones.

CONNECTIVE TISSUE

·

- **Cellular Components**
- **Fibrous Components**
- **Ground Substance**
- **Basement Membrane**
- **Histologic Organization of Connective Tissue**

·   ·   ·   ·   ·   ·   ·   ·   ·   ·   ·   ·   ·

Connective tissue (CT) is what is left over after everything else (epithelia, muscle, and nervous tissue) is removed. As its name implies, it connects tissues to one another (e.g., muscle to bone) and serves as a packing between tissues. CT is found in every organ system except the CNS where its role is served by glial cells, particularly astrocytes. CT surrounds muscles, blood vessels, nerves, and organs, and it underlies and anchors epithelia. Bone, cartilage, and blood are specialized connective tissues and are discussed in subsequent chapters.

CELLULAR COMPONENTS

Fibroblasts are the principal cell type found in CT.

Fibroblasts, derived from embryonic mesenchyme, are the most numerous cells found in CT. These spindle-shaped cells containing elongated, condensed nuclei secrete the majority of fibrous proteins (collagen and elastin) and ground substance (the sugar-based molecules, glycosaminoglycans and proteoglycans) which make up the extracellular matrix (ECM). In most H&E-stained sections, fibroblasts can be identified by their dark, wavy nuclei, the cytoplasm being too attenuated to be visible.

Other cells types found in CT are derived either from mesenchymal precursors or from white blood cell lineages (bone marrow–derived).

Fat cells (*adipocytes*) are derived from undifferentiated mesenchymal cells and are distinguished by their unique appearance. White adipocytes are very large (up to 150 μm), with most of the cellular volume occupied by a single fat droplet (*unilocular*). The nucleus is pushed to the periphery, and cytoplasm is almost absent, giving these cells the classic "signet ring" appearance. White fat is distributed throughout the body. Brown fat cells, on the other hand, contain numerous smaller lipid droplets (*multilocular*), slightly more cytoplasm, and abundant mitochondria which give them their brown coloration. Brown fat is found in fetuses and neonates and only in very restricted areas in adult humans. It is more prevalent in hibernating mammals and provides heat through a process which bypasses the production of adenosine triphosphate (ATP) from the chemiosmotic gradient in mitochondria, the energy instead being released as heat (nonshivering thermogenesis).

The white blood cell derivatives found in CT include mast cells (and their cousins, the basophils), macrophages, plasma cells, lymphocytes, neutrophils, and eosinophils. *Mast cells* are most easily identified in sections of tissue stained with toluidine blue by their numerous, large, metachromatic (color-changing) secretory granules containing *histamine and heparin*. Mast cells respond to foreign substances (allergens) by rapid secretion of their granular contents via the regulated secretory pathway (Chap. 6), resulting in the *immediate hypersensitivity response* characteristic of allergic rhinitis (the sniffles).

Lymphocytes are identifiable by their dark-staining round nuclei and scanty cytoplasm. These clues denote that these cells are not actively transcribing mRNA and not actively synthesizing protein. In contrast, *plasma cells* (activated B lymphocytes) are characterized by large, pale-staining nuclei (euchromatin) with a clock-face distribution of heterochromatin, denoting active transcription of immunoglobulin mRNA. A perinuclear light-staining region of cytoplasm indicates a prominent Golgi apparatus involved in the modification and routing of newly synthesized *immunoglobulins*. (See Chap. 4 for a visual comparison.)

Eosinophils, characterized by their red-staining (eosinophilic) storage vesicles in H&E preparations, are phagocytes lured from the blood into CT by mast cell secretions [heparin and eosinophil chemotactic factor of anaphylaxis (ECF-A)] and limit the effects of other mast cell secretions. *Macrophages* are derived from monocytes. They phagocytose bacteria and cellular debris and present modified antigens to lymphocytes. *Neutrophils* are also phagocytes but are less active than macrophages. They possess secretory granules containing various enzymes which, when secreted, result in the dissolution of surrounding tissue components. Neutrophils are the initial responders in inflammatory reactions and are the major cellular component of pus. Movement of these cells from the blood into the tissues is discussed in Chap. 18, whereas their roles in the process of inflammation are covered in Chap. 19.

Components of the ECM are secreted constitutively by fibroblasts.

In contrast to the secretion of products stored in vesicles and released upon appropriate stimuli, the secretory products of fibroblasts (fibrous proteins and sugar-based ground substance) are shuttled directly and continuously from the trans-Golgi network (TGN) to the cell membrane in coated vesicles (Chap. 6).

FIBROUS COMPONENTS

Collagen type I and elastin are the predominant fibrillar proteins secreted by fibroblasts.

Type I collagen is synthesized as a prepropeptide, the signal (pre-) sequence being cleaved after translocation of the polypeptide into the endoplasmic reticulum (ER) lumen. The pro-collagen molecule is assembled from three α-chains organized into a triple helix (Fig. 13-1). Amino- and carboxy-terminal extensions prevent procollagen from forming collagen fibers within the ER, Golgi apparatus, and transport vesicles. Upon secretion, peptidases cleave the nonhelical terminal peptides resulting in *tropocollagen*, which spontaneously assembles in staggered arrays to form collagen fibers with a characteristic 64-nm banding pattern observed in electron micrographs. Collagen contains high concentrations of hydroxyproline and hydroxylysine which stabilize the triple helices and cross-link tropocollagen monomers, respectively.

Elastin is also synthesized as a prepropeptide, secreted as a propeptide which is converted to tropoelastin by extracellular enzymatic cleavage, and subsequently assembled into amorphous fibers or sheets with the aid of fibrillin, a fibrous glycoprotein also secreted by fibroblasts. Elastin allows deformed tissues to return to their original shapes. Elastin is best viewed at the light microscopic level as large bundles or sheets between layers of smooth muscle cells in the walls of large "elastic" arteries such as the aorta. In this case, the elastin is secreted by the smooth muscle cells, not by fibroblasts.

Collagen exists in forms (at least 19) which are primarily dictated by the types of α-chains compiled to form the triple helix. Many of these collagens are site or tissue specific. The most well characterized collagen types are listed in the table.

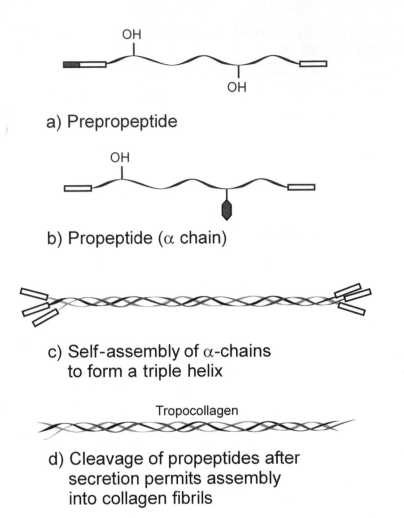

a) Prepropeptide

b) Propeptide (α chain)

c) Self-assembly of α-chains
to form a triple helix

Tropocollagen

d) Cleavage of propeptides after
secretion permits assembly
into collagen fibrils

Figure 13-1 Synthesis of collagen.
Collagen is synthesized as a prepropeptide containing lysine and proline residues, some of
which are enzymatically hydroxylated (a). During subsequent processing in the RER, the
signal sequence (shaded box) is removed and some hydroxylysine residues may be glycosylated (b). The resulting α chains containing nonhelical portions at both ends (open
boxes) self-assemble into a triple helix (c). Following secretion, enzymatic removal of the
nonhelical ends permits assembly of tropocollagen into collagen fibrils (d).

MOST WELL-CHARACTERIZED TYPES OF COLLAGEN

TYPE	LOCATION	FUNCTION
I	General CT and bone	Tensile strength
II	Hyaline and elastic cartilage	Tensile strength
III	Parenchyma of organs and walls of blood vessels	Reticular framework
IV	Basement membranes	Meshwork scaffolding
V	Muscle basal lamina	Meshwork scaffolding, unknown
VI	Ubiquitous	Unknown
VII	Basement membrane of skin and amnion	Anchoring fibers
VIII	Endothelium	Unknown
IX–XII	Cartilage	Unknown

Types I to III and others are regarded as the *fibrillar collagens* since they form banded fibers. They function in a similar fashion to steel "rebars" used in construction. In contrast, type IV and type VII *(network-forming) collagens* found in epithelial basal laminae form a lattice-like network similar to a chain-link fence or chicken wire. Type IX-type XII collagens are known as the *FACIT collagens (fibril associated collagens with interrupted triple helices)*. They regulate the orientation and function of the fibrillar collagens, IX being associated with type II collagen and XII associated with type I collagen, respectively. Unlike the fibrillar collagens, the FACIT collagens retain their terminal registration peptides.

GROUND SUBSTANCE

Glycosaminoglycans (GAGs) regulate the water content and effective pore space through which substances diffuse in the ECM.

Glycosaminoglycans are large, negatively charged polysaccharides consisting of repeated disaccharide units. They are typically rigid structures and thus maintain large, water-filled spaces in the ECM. In addition, the negative charge binds Na^+, which enhances water retention (turgor).

GAGs fall into four groups with different but overlapping distributions: (1) hyaluronic acid (cartilage, skin, synovial fluid, and general CT); (2) heparin and heparan sulfate (basement membrane, skin, lung, liver, blood vessels, and mast cell granules); (3) chondroitin sulfate and dermatan sulfate (cartilage, bone, skin,

blood vessels, heart, cornea); and (4) keratan sulfate (cartilage, cornea, and inter-
vertebral disk).

GAGs attach covalently to proteins to form proteoglycans.

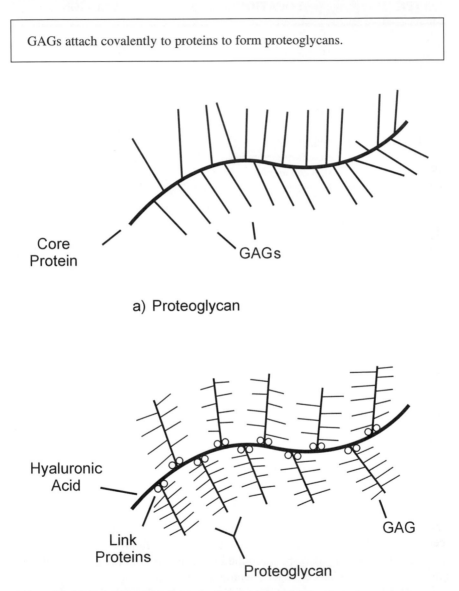

Figure 13-2 Proteoglycans and proteoglycan aggregates.
Glycosaminoglycans (GAGs) consisting of repeated disaccharide units linked to core pro-
teins to form proteoglycans (*upper figure*). Proteoglycans attach to hyaluronic acid (also a
GAG) with the help of link proteins to form proteoglycan aggregates (*lower figure*).

With the exception of hyaluronic acid, GAGs covalently link to core proteins to form *proteoglycans* (Fig. 13-2). The type of GAG linking to a core protein determines the specificity of pore size and charge of the proteoglycan. Hyaluronic acid links proteoglycans into larger aggregates. Proteoglycans are primarily responsible for maintaining the water-filled space within the ECM.

Glycoproteins link other components within the ECM and anchor cells to the ECM.

Fibronectin is a fibrillar glycoprotein which plays a major role in anchoring cells to the ECM. It contains binding sites for its cellular receptor, a member of the integrin family of transmembrane proteins, as well as for collagen fibers of the ECM. Nonfibrillar glycoproteins also play a role in binding components of the ECM to each other including *laminin, entactin, and tenascin.* Laminin, with entactin, links components of the basal lamina, including type IV collagen and GAGs (heparan sulfate), to integrins. These ECM interactions are fully described in Chap. 8.

BASEMENT MEMBRANE

Basement membranes are an interface between epithelial, muscle, nerve, fat, or Schwann cells and the underlying supportive connective tissue.

The basement membrane is composed of two parts: (1) the basal lamina (external lamina in nonepithelial cells), which is the portion synthesized by the epithelium, consisting of a lamina densa containing type IV collagen and heparan sulfate, and variable laminae rarae (GAGs and proteoglycans, particularly laminin and entactin) visible only with the electron microscope and (2) a reticular lamina of small collagen fibrils (type III). Type VII collagen anchors the basal lamina and reticular lamina to underlying ECM at the location of hemidesmosomes (see Chap. 7). In addition, cells are attached to fibronectin in the basal lamina by integrins (fibronectin receptors). Fibronectin contains binding sites for cells, heparan sulfate, and collagen.

The basement membrane serves as a molecular sieve and plays a role in cellular organization, differentiation, and migration in addition to its anchoring function.

The basement membrane serves as a barrier to the permeability of large molecules, particularly proteins. The effective size of its "pores" is dependent on the charge and spatial distribution of its component GAGs. The glomerular basement membrane of the kidney is an excellent example (see Chap. 24).

As previously noted, the basement membrane serves as an attachment site where cells may be anchored to the underlying connective tissue matrix via several mechanisms such as integrins and fibronectin. Interaction of cell surface receptors with components of the basal lamina induces noncommitted cells to enter specific differentiation pathways, provides specific routes for migration of cells during development, and plays a role in inhibiting the spread (metastasis) of tumor cells. The penetration of blood-borne cells through vascular endothelium and its associated basement membrane is discussed in Chap. 18.

HISTOLOGIC ORGANIZATION OF CONNECTIVE TISSUE

General connective tissue is divided histologically into three categories:

1. Loose CT is characterized by a high ratio of cells (fibroblasts) to fibrous components (primarily type I collagen) of the ECM. It is usually found immediately beneath the epithelia of most organ systems.
2. Dense irregular CT (DICT) is characterized by a high ratio of fibrous components to fibroblasts. The collagen tends to be present in thicker bundles than those found in loose CT. The bundles have no specific orientation. The DICT links the overlying loose CT and epithelium to deeper structures.
3. Dense regular CT contains thick, highly ordered collagen bundles with few inactive fibroblasts (fibrocytes). In tendons and ligaments, the fibers and cells tend to be oriented along the axis of tension.

SPECIALIZED CONNECTIVE TISSUES: CARTILAGE, BONE, BONE DEVELOPMENT, AND FRACTURE HEALING

·

- **Cartilage**

- **Bone**

- **Constituents of Bone**

- **Osteogenesis**

- **Regulation of Osteoblastic and Osteoclastic Activity**

- **Bone Repair (Fracture Healing)**

· · · · · · · · · · · ·

Cartilage and bone are specialized connective tissues. They are classified as connective tissue because they consist of cells, fibers, and ground substance (fibers + ground substance = matrix). They are skeletal tissues as well, providing support and protection for the soft tissues and organs of the body while allowing flexibility. Cartilage is more flexible than bone, which contains mineral deposits limiting its ability to bend or flex. Bone provides a vast Ca^{2+} reservoir required for most bodily functions.

CARTILAGE

Cartilage consists of cells (chondrocytes) and extracellular fibers in a ground substance. The ground substance and fibers together constitute the matrix. There are three subtypes of cartilage: (1) *hyaline*, most common, contains collagen fibers; (2) *elastic*, specialized by addition of elastic fibers to the matrix; and (3) *fibrous (fibrocartilage)*, increased collagen in matrix, reduced cellularity compared to hyaline cartilage (see table). The *perichondrium* is the envelope of connective tissue which surrounds hyaline and elastic cartilage.

FEATURES OF THE DIFFERENT TYPES OF CARTILAGE

CARTILAGE TYPE	OCCURRENCE	MACROSCOPIC APPEARANCE	MICROSCOPIC APPEARANCE	PRESENCE OF PERICHONDRIUM
Hyaline	External auditory meatus, larynx, tracheal rings, bronchi, fetal long bones, articular ends of long bones	Translucent, bluish gray, solid but flexible	Chondrocytes in cell groups (resulting from cell division), type II collagen fibers	Yes
Elastic	Auricle of the ear (pinna), epiglottis	Yellow color (presence of many elastic fibers), and more opaque, flexible, and elastic than hyaline cartilage	Chondrocytes mostly located singly, type II collagen plus elastic fibers	Yes
Fibro-cartilage	Intervertebral disks and pubic symphysis; insertion of some tendons and ligaments; closely associated with dense c.t. or hyaline cartilage	Opaque appearance, firm, fibrous texture	Type I collagen, single, sparse chondrocytes	No

The arrangement of the hyaline cartilage ground substance is shown in Fig. 14-1. The proteoglycans consist of a protein core with polysaccharide chains of more than one type attached (i.e., keratan sulfate or chondroitin sulfate).

> The biomechanical characteristics of cartilage are based on the molecular arrangement of its constituents.

For example, the side chains of the glycosaminoglycans (GAGs) are negatively charged resulting in repulsion (Fig. 14-1) and wide spacing which is usually filled with water. Water associates with the negative charge resulting in a shock absorbing affect. This is an important characteristic for the hyaline cartilage, which covers articular surfaces. In that case, water enters the cartilage when pressure is relieved. When pressure is applied to the joint, water is released from the cartilage and enters the cavity of the joint. Another example of the molecular basis of tissue characteristics is seen in the case of the rigidity or flexibility of

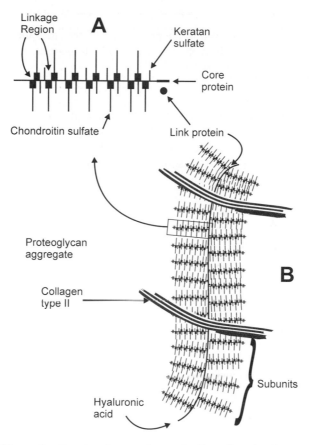

Figure 14-1 The molecular structure and organization of proteoglycans.
Proteoglycans (*A*) are the subunits that are noncovalently associated with hyaluronic acid
to form aggregates (*B*). The proteoglycans consist of a protein core with glycosaminoglycans (polysaccharide chains) of more than one type attached (e.g., keratan sulfate and
chondroitin sulfate) by covalent linkage. Cross-linking between collagen type II and proteoglycans is responsible for the rigidity of the cartilage matrix.

hyaline cartilage, which is largely based on the cross-linking between collagen
and the proteoglycans (see Fig. 14-1).

Synthesis of matrix components

Cartilage ground substance contains a high concentration of *chondroitic acids*.
The chondroitic acids are the polysaccharides, chondroitin 4-sulfate (chondroitin
A) and chondroitin 6-sulfate (chondroitin C), which are responsible for the
basophilia of the matrix observed around groups of cells (capsular or territorial
matrix).

The chondrocyte is involved in the production of collagen and proteoglycans of the matrix as well as chondronectin, which increases adhesiveness of chondrocytes to the matrix.

Amino acids are incorporated into protein on the ribosomes. Sugars and sulfates are incorporated into polysaccharide in the Golgi region and combined to form the protein-polysaccharide molecules (proteoglycans) released constitutively along with collagen into the surrounding matrix.

Nutrition

Cartilage is avascular, which has dramatic implications for nutrition, growth, and repair.

Cartilage lacks intrinsic blood vessels, nerves, and lymphatics. Nutrients, oxygen, and cell waste must seep through the matrix by diffusion from the perichondrium (surrounding sheath of connective tissue). The avascularity of cartilage results in a low rate of metabolism in adult cartilage as well as delayed or inhibited responses to damage and injury.

Histogenesis

Centers of chondrification consist of mesenchymal cell aggregations that differentiate into chondrocytes which secrete collagen, ground substance, and chondronectin. The mesenchyme surrounding the original site becomes compressed and is surrounded by the dense connective tissue of the perichondrium. The perichondrium brings blood vessels into close association with cartilage and provides a source of new cells during development.

Subsequent cartilage growth occurs by interstitial and appositional mechanisms.

Interstitial growth (growth from within)

Isogenous groups of cells (cell nests) are formed by mitosis combined with increased deposition of matrix as the progeny differentiate and synthesize fibers and ground substance. This leads to an *expansion of the cartilage from within*.

Appositional growth (growth from the perichondrium)

The innermost cells of the perichondrium differentiate into immature chondrocytes and secrete matrix. These cells become overlaid by newer cells (derived from the perichondrium) and their matrix. Interstitial and appositional growth occur in hyaline and elastic cartilage. Growth of fibrocartilage is more like connective tissue due to the absence of a perichondrium.

Regeneration and repair of cartilage in adult mammals

There is only a minimal amount of cartilage regeneration and repair in adults, and it occurs through the limited activity of immature chondrocytes originating from the perichondrium. At times regeneration is limited to infiltration of granulation tissue to form a connective tissue scar. The requirement of diffusion for movement of growth factors and nutrients combined with a low metabolic rate explains the slow repair which occurs in cartilage following injury.

BONE

Bone, like cartilage, is a specialized connective tissue containing cells, fibers, and ground substance. In comparison to cartilage, bone has some unique qualities. The deposition of lime salts, in the form of *hydroxyapatite* crystals, prevents diffusion of metabolites and interstitial growth, both of which occur in cartilage.

The distinctive features of bone are as follows: (1) a canalicular system, (2) vascularity, (3) elongation through appositional growth of a hyaline cartilage model, and (4) continuous resorption, reconstruction, and remodeling.

Bone contains tiny canals extending from one cell to another and opening into tissue spaces, allowing close proximity to the blood supply. *Unlike cartilage, which is avascular, bone is highly vascular;* bone cells are found in close proximity to capillaries. Elongation of bone occurs by appositional type growth. Interstitial growth as occurs in cartilage is impossible because of the presence of hydroxyapatite crystals.

Bone architecture is not static.

Fetal bone is destroyed locally and reformed, with the process recurring many times as changes in soft tissue are transduced to the calcified tissue. After birth and in subsequent reconstructions, parallel systems of *lamellae (Haversian systems)* are laid down. The Haversian system, or *osteon* (Fig. 14-5), is formed by concentric lamellae, or rings, comprising bone forming cells and their surrounding bony matrix. The rings represent progressive layers of deposition and are built around centrally positioned blood vessels. Reconstruction of lamellae through bone deposition and resorption occurs continuously throughout life resulting in multiple generations of lamellae.

Macroscopic structure of bone

> There are two types of bone in adults: lamellar (compact) bone and trabecular (cancellous) bone.

Compact bone lacks cavities and consists of a dense plate on the outside of long bones or flat bones (Fig. 14-2). Trabecular bone has a three-dimensional lattice of branching, bony spicules intertwined to form trabeculae surrounding the bone marrow spaces in the long bones and flat bones (Fig. 14-2). There are differences in trabecular and compact bone growth and maintenance as well as loss of mineralization with advancing age.

The geographic regions of the long bone are shown in Fig. 14-2. The diaphysis, or shaft, is the thick-walled hollow cylinder of compact bone with a central medullary (marrow) cavity. The epiphyses are the sites of growth after birth in long bones and contain the 2° ossification centers or epiphyseal plates. The metaphysis is a transitional region between the epiphysis and diaphysis. The articular cartilage is composed of hyaline cartilage covering compact bone at joint (articular) surfaces. The articular cartilage lacks a perichondrium.

General features of bone

Bone is normally surrounded externally by a layer of specialized dense connective tissue (*periosteum*) similar to the perichondrium of hyaline cartilage. The periosteum has osteogenic potential and is absent in several regions: (1) articulating surfaces and (2) attachment of tendons and ligaments. Lining the bony surfaces of the medullary spaces, as well as spaces in cortical bone traversed by blood vessels, is a layer of attenuated connective tissue cells with osteogenic potential; this layer is known as the *endosteum*. In the skull, the periosteum is known as the *pericranium*; the endosteum is referred to as the *endocranium*. The outer layer of the dura mater (the layer of the meninges closest to the skull) func-

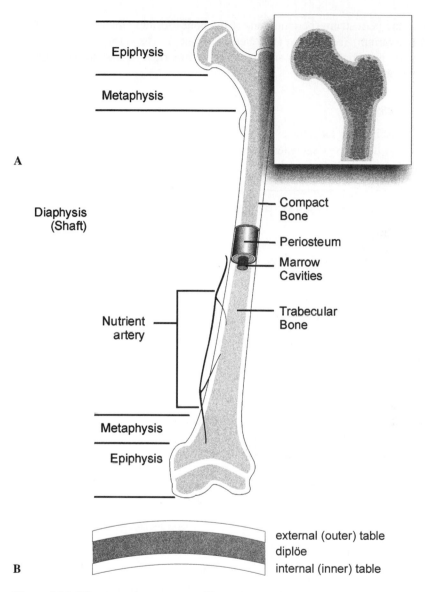

Epiphysis

Metaphysis

A

Diaphysis
(Shaft)

Compact
Bone

Periosteum

Marrow
Cavities

Nutrient
artery

Trabecular
Bone

Metaphysis

Epiphysis

external (outer) table
diplöe
B   internal (inner) table

Figure 14-2 Macroscopic structure of bone.
A. A section through the length of a long bone such as the femur or humerus illustrating
the areas of compact and cancellous bone. *B.* A section through a flat bone of the skull
such as the parietal or frontal bone illustrating the regions of compact and cancellous bone.
The two zones of compact bone are called tables, whereas the area in between the tables
is known as the diplöe.

tions as the periosteum (pericranium) of the bones forming the brain case of the skull (calvarium) (see Chap. 16, "Neural Tissue"). The dura mater maintains osteogenic potential in adults and therefore is important in fracture healing and bone remodeling.

Blood vessels

Blood vessels enter long bones through nutrient foramina or canals (periosteal buds) derived from vessels entering the epiphyses. Nerves (sympathetic and pain fibers) accompany these vessels.

CONSTITUENTS OF BONE

Bone cells

Bone cells are derived from two separate cell lineages.

Osteoprogenitor cells are stem cells which in the adult are described as the *bone lining cells* (Fig. 14-3). In the adult, they are found in the inner portion of the periosteum, in the endosteum, and lining vascular canals of compact bone. These cells are derived from mesenchyme of the embryonic somite (specifically the sclerotome) and possess mitotic potential.

Osteogenitor cells give rise to the osteoblast/osteocyte lineage.

Osteoblasts synthesize and secrete the organic matrix of bone (collagenous and noncollagenous proteins) and alkaline phosphatase. Cellular polarity is apparent, and there is a resemblance to other secretory cells. Extensive rough

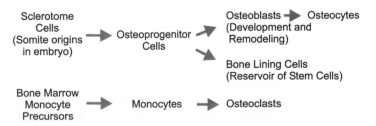

Figure 14-3 The separate lineages and embryologic origin of osteoblasts and osteocytes compared with osteoclasts.
Osteoclasts are derived from the bone marrow, whereas osteoblasts and osteocytes are derived from the paraxial mesoderm, specifically the sclerotome of the somite.

endoplasmic retriculum (RER) is clustered in the cytoplasm nearest the bone surface. Bone-specific alkaline phosphatase has been detected on the cell membrane of osteoblasts. Golgi and secretory vesicles are common in osteoblasts.

Osteocytes are derived from osteoblasts which become trapped by the matrix they secrete. They are located in potential spaces called *lacunae* ("lakes"). In life these lacunae are filled with bone fluid (Fig. 14-4). The processes of the osteoblasts are surrounded by bone in slender canals (canaliculi) as they mature. Electron microscopy (EM) reveals diminished RER and Golgi compared with an osteoblast, but the cells are functionally active in maintaining bone and play an important role in mineral storage and possibly minimal amounts of resorption.

Osteoclasts are derived from monocytes and are part of the monocyte-phagocyte system.

Osteoclasts are giant multinucleate, motile cells derived from *monocytes*. They possess a foamy, eosinophilic cytoplasm with numerous lysosomes, vesicles, and vacuoles containing acid hydrolases, which produce an appropriate acidic environment for breakdown of the bony matrix. These enzymes are targeted for secretion by the presence of the mannose-6-phosphate receptor. The plasmalemma of the osteoclast adjacent to the resorbing bone surface is thrown into folds and villous-like processes whose tips reach and even enter the bone surface. This surface is called the *ruffled border*. The osteoclast is attached to the

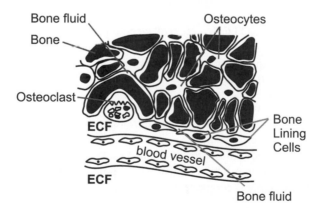

Figure 14-4 The relationship between the bone fluid, extracellular fluid, bone matrix, osteocytes, osteoclasts, and bone lining cells.
Remember that when an osteoclast resorbs bone and releases Ca^{2+} into the bone fluid, the bone fluid is continuous with the extracellular fluid. The overall result of osteoclastic activity is an increase in blood Ca^{2+} levels.

bone surface, and the resorption area is sealed off by the presence of contractile proteins in the cytoplasm. The basolateral membrane of the osteoclast has a Na^+, K^+-ATPase pump.

> The osteoclast does not resorb bone like a vacuum cleaner, but acidifies its local environment.

The *osteoclast* acidifies the bone compartment by pumping protons manufactured in the cytosol by action of carbonic anhydrase. The bone compartment around an osteoclast, although extracellular in location, is therefore analogous to a secondary lysosome (Chap. 6) since it contains (1) an acidic pH (for dissolving hydroxyapatite crystals), (2) lysosomal enzymes, and (3) the enzymatic substrate.

Bone matrix proteins

Osteoblasts synthesize collagen, noncollagenous proteins, and *alkaline phosphatase*. Newly synthesized nonmineralized matrix is known as *osteoid*, or prebone, and is calcified after secretion of alkaline phosphatase. The following paragraphs explain the constituents of bone.

• **COLLAGEN**   Type I collagen makes up 85 to 90 percent of total bone protein. It differs from type I collagen found elsewhere in the body because of posttranslational modifications which are believed to facilitate the movement of hydroxyapatite crystals into the fibers to induce nucleation and growth.

• **NONCOLLAGEN PROTEIN**   Noncollagenous proteins make up 10 to 15 percent of bone protein content. About 25 percent of the noncollagenous protein is derived exogenously and includes trapped growth factors. The noncollagenous proteins include the following:

- Cell attachment proteins (fibronectin, *osteopontin*)
- Proteoglycans (i.e., chondroitin 4- and chondroitin 6-sulfate, which may play a role in collagen fibrillogenesis)
- Gamma-carboxylated (Gla) protein-gamma-carboxyglutamic acid residues are found in osteocalcin which is also called bone Gla protein. *Osteocalcin* binds Ca^{2+} and mineral components of the matrix. It is used as *a marker of bone turnover.*
- Growth-related peptides [transforming growth factor-β, insulin-like growth factor-I, bone morphogenetic protein (BMP), osteonectin].

• **BONE (MINERAL) SALTS**   Calcium, phosphate, and hydroxyl ions form a calcium phosphate complex in bone tissue called *hydroxyapatite*:

$Ca_{10} (PO_4)_6(OH)_2$. Hydroxyapatite crystals are deposited on collagen fibers and may contain: Na^+, Mg^{2+}, and F^- as well as citrate, carbonate, and other ions.

Arrangement of constituents

There are two principal ways to histologically prepare bone. Decalcified sections contain organic materials such as cells, matrix, and blood vessels. Ground bone sections contain only bone. All cellular and organic materials are absent, their former locations represented histologically as blackened spaces caused by entrapped air.

Lamellar or compact bone consists of concentric (Haversian) lamellae which encircle a central blood vessel (and associated nerves) forming an osteon, or Haversian system. Osteocytes within their lacunae are arranged in the osteons similar to planetary "static" orbits around a centrally-located sun like our solar system. The bone surrounding the osteocyte processes form canaliculi radiating towards the central (Haversian) canal. Adjacent Haversian canals may be connected with each other, the external bone surface, or the marrow cavity by channels called Volkmann's canals, running approximately perpendicular to their planes (Fig. 14-5).

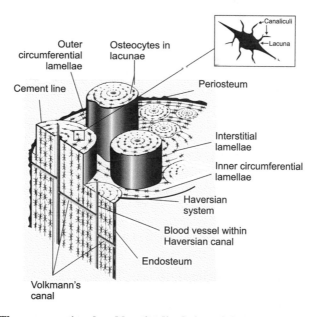

Figure 14-5 The cross-sectional and longitudinal view of the structural arrangement of compact bone.
The Haversian system, the periosteum, and the endosteum are interconnected by Volkmann's canals.

OSTEOGENESIS

Osteogenesis occurs in only one way: by the action of osteoblasts.

Osteoblasts synthesize fibers and ground substance as well as alkaline phosphatase, which induces matrix calcification. Osteogenic activity is soon accompanied by the selective destructive action of osteoclasts in a remodeling process, continuously adapting the growth of bone to developing soft tissues and dynamic mechanical forces. Growth by remodeling is necessary, because interstitial growth by bone cells cannot take place.

Two types of bone formation are noted, dependent on whether bone is formed *de novo* in a soft tissue area (intramembranous ossification) or formed in a site occupied by an established cartilaginous model (endochondral ossification).

General aspects of osteogenesis (bone formation)

Bone first appears as little spikes called *spicules*. Osteoblasts form a matrix consisting of *type I collagen* and an amorphous ground substance. This provisional matrix is called *osteoid* and is later stiffened by *hydroxyapatite* to form the definitive matrix. The deposition of matrix is incomplete around each bone cell, resulting in *lacunae*, and a tubular *canaliculus* surrounds each cell process (see Fig. 14-5). The calcification process is dependent on the action of alkaline phosphatase in releasing PO_4^{2-} ions to stimulate the formation of $Ca_3 (PO_4)_2$.

Woven bone is the first formed bone in any type of bone formation. The term *woven* refers to the embryonic condition, in which fibers and cells are randomly oriented. Woven bone results from the initial branching and elongation of spicules to form trabeculae and consists of tortuous channels of bone surrounding blood vessels. Later in development, the woven bone of the fetal skeleton is replaced by trabecular and compact bone in a remodeling process.

Intramembranous ossification

In intramembranous ossification, bone cells form directly from mesenchyme (membrane).

Intramembranous ossification occurs in the flat bones of the cranial vault and irregular bones of the face. Some bones are mixed membrane and cartilage bones: occipital, temporal, and sphenoid.

In the primary ossification center, mesenchymal cells differentiate into the osteoblast lineage (osteoprogenitor cells → osteoblasts) and start to lay down branching spicules and trabeculae of bone. Preceding the differentiation of osteoblasts, the mesenchyme becomes highly vascularized. The surrounding connective tissue is organized as a *periosteum*.

Collagen and noncollagenous proteins are secreted by the osteoblasts, and they polymerize to form randomly interwoven fibrils of collagen throughout the trabeculae of osseous matrix. The osteocytes are distributed uniformly, but oriented randomly. The woven bone is remodeled primarily after birth in the skull vault to give two denser plates of bone, or *tables (compact bone)*, with a spongy bone marrow space between the tables of *trabecular bone (diplöe*; Fig. 14-2). The plates expand from their primary centers, but during the growth period remain separated by connective tissue *sutures*, permitting outward growth to accommodate the brain. These sutures and the *fontanelles* (intersection of two or more sutures) are designed to permit growth and expansion of the skull and brain.

Endochondral ossification

In endochondral ossification, a cartilage model of the long bone is formed and replaced by bone.

Mesenchymal cells differentiate into chondrocytes and form a miniature hyaline cartilage model of approximately the shape of the eventual skeletal bone such as the femur. This cartilage model has a perichondrium (Fig. 14-6). As bone development begins, the following events occur in the central region of the shaft (diaphysis):

1. Cartilage cells enlarge or hypertrophy (the chondrocytes accumulate glycogen, and the cytoplasm becomes vacuolated); the hypertrophied chondrocytes synthesize alkaline phosphatase.
2. The matrix around the hypertrophied chondrocytes becomes calcified in response to the action of alkaline phosphatase.
3. A vascular-osteogenic bud of osteoblasts, endothelial cells, osteoclasts, macrophages, and bone marrow cells invades the calcified matrix. (Steps 3 and 4 are interdependent.)
4. The perichondrium in the area of the vascular bud is converted to a periosteum, since this layer initiates osteoblast differentiation instead of chondro-

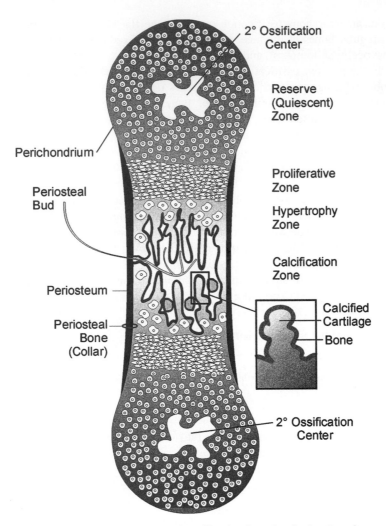

Figure 14-6 The zones of endochondral ossification in a developing long bone.
The zones are shown for the primary ossification center of a long bone during fetal development. Similar zones develop in the secondary ossification centers after birth.

cyte differentiation. Bone is laid down under the periosteum to form a *bony collar*, or *periosteal collar*. Actually, this bone is formed by intramembranous ossificiation since there is no cartilage intermediate in the periosteal region.

5. The primary ossification center now is made up of bone forming in, and on, the calcified walls of spaces eroded in the cartilage matrix. The calcified cartilage forms a scaffolding for bone deposition.

Growth and formation of long bones

A primary ossification zone establishes itself across the width of the shaft and begins extending in both directions toward the epiphyses, resulting in two fronts of primary ossification extending across the diaphysis (Fig. 14-6). Between the fronts and epiphyses are *zones of endochondral ossification* as follows:

1. Reserve or *quiescent zone* of hyaline cartilage, in which little growth and cell division occur. This zone is farthest from the mineralization front.
2. *Proliferative zone*, in which cartilage cells multiply and arrange themselves in ordered parallel columns of disk-shaped cells often compared to the appearance of a roll of dimes viewed from the side. Growth in the long axis of the bone occurs primarily through the formation of new cells here.

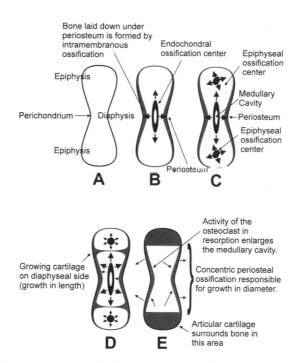

Figure 14-7 The changes that occur in a long bone in transformation from a cartilage model to a fully formed long bone.
A. The cartilage model. B. The location of the primary ossification center. C. The position of the primary and secondary (epiphyseal) ossification centers. E. The formation of the marrow cavities. Remember, the important concepts of bone development: (1) Growth in length occurs by proliferation of the chondrocytes in the primary and/or secondary ossification center depending on the stage of fetal or postnatal development and (2) Growth in diameter occurs by deposition of new bone under the periosteum.

3. *Zone of hypertrophy*, in which the chondrocyte columns have halted mitosis and become enlarged. These chondrocytes produce alkaline phosphatase.
4. *Calcification zone*, in which the matrix around the hypertrophied cells becomes impregnated by calcium phosphate salts (hydroxyapatite crystals).
5. *Zone of cartilage removal and bone deposition*, in which degeneration and death of cartilage cells occurs due to calcification of the matrix which the chondrocytes depend on for nutrients. New bone is deposited on the calcified cartilage.

Epiphyseal (secondary) ossification centers

Within the cartilage of the epiphysis, a secondary ossification center later develops, again with processes of cartilage cell hypertrophy, matrix calcification, and its erosion by vascular and other elements penetrating from the perichondrium. However, tall, orderly columns of chondrocytes and a defined marrow cavity are not seen. The epiphyseal, secondary ossification center spreads to occupy much of the epiphysis and to form another bony border to the cartilaginous epiphyseal plate. Although the cartilage in the plate can grow (thus lengthening the whole bone), it cannot keep pace with the front of ossification invading it from the diaphyseal side. Eventually, ossification overtakes cartilage proliferation, and the primary ossification front fuses with the secondary epiphyseal bone. The *epiphyseal plate* is obliterated, and longitudinal growth ceases. This is known as *epiphyseal closure* (or *closure of the 2° ossification center*), and a dense line of bone marks its site. Hyaline cartilage remains as a thin cap over the epiphysis to form an articulating surface. Epiphyseal closure takes place at different times in different bones and can be used in determining the approximate age of a specimen.

The marrow cavity in the shaft develops from the small primary spaces in the trabeculae of bone which are first formed. It forms by erosion of the ends of the trabeculae and of the inner surface of the shaft sides to become a well-defined space.

Growth in length occurs through the proliferation of chondrocytes; increase in width occurs through periosteal deposition.

In the fetus, growth in length of long bones occurs by proliferation of the chondrocytes in the proliferative zone of the primary ossification center. After birth, growth in length occurs by proliferation of chondrocytes in the epiphyseal disk (cartilage) situated between the epiphysis and diaphysis. Growth in width of a long bone occurs by addition of new bone under the periosteum along with simultaneous osteoclastic resorption (enlarging the medullary cavity, Fig. 14-8).

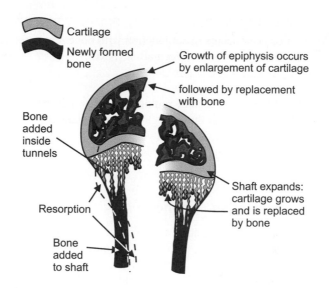

Figure 14-8 The deposition and resorption of bone, which results in the remodeling of the shape of the long bones of the body.
The structures which appear as trabeculae in the longitudinal section are slices that have been cut through the walls of tunnels. The trabeculae consist of spicules composed of cores of calcified cartilage surrounded by bone. (*Modified from Cormack: Ham's Histology, 9th Edition, J. B. Lippincott, 1987.*)

> Bone resorption and bone deposition are tightly linked both during development (osteogenesis) and in the adult during bone remodeling.

In the case of both growth in length and width, bone deposition on one surface is counteracted by bone resorption on the opposite surface. The balance between bone deposition and resorption is a homeostatic mechanism which goes awry in pathologic conditions involving bone. Bone resorption and bone deposition are closely linked in a cycle of events.

Activation-resorption-reversal-formation cycle

The inexorable linkage of bone deposition and resorption is seen at the systems level (hormone regulation), local level (growth factors), and cellular level (osteoblast-osteoclast interactions).

Bone deposition only occurs in conjunction with bone resorption in a cycle of activation-resorption-reversal-formation (ARF).

The ARF cycle occurs at adult remodeling sites, but also during development. The ARF cycle explains the linkage between osteoclastic and osteoblastic activity. During development there are two ARF cycles, one to replace calcified cartilage with woven bone and one to replace woven bone with lamellar (Haversian) bone.

Internal reorganization of bone

The original pattern of endochondral ossification in the middle of the diaphysis and intramembranous bone along the wall of the diaphysis is erased by continual internal resorption and reorganization of bone in most mammals, including humans. In the manatee (sea cow) and in children with some forms of inherited *osteopetrosis*, there is no secondary marrow cavity, and the contributions of subperiosteal and endochondral bone formation in the shaft of the long bone remain distinct. Osteopetrosis is a disease in which dense bone is formed because of the failure of normal resorptive processes. In the childhood form of osteopetrosis mentioned above, there is a deficiency in carbonic anhydrase type II isoenzyme, which results in the presence of incompetent osteoclasts and the absence of a marrow cavity.

REGULATION OF OSTEOBLASTIC AND OSTEOCLASTIC ACTIVITY

Hormones, growth factors, cytokines, and vitamins regulate bone development and maintenance.

The secretion of osteoblasts is regulated by a number of factors. Some of these factors also regulate bone resorption, which should not be surprising considering the linkage of the ARF cycle.

Parathyroid hormone (PTH) is the primary regulator of bone turnover.

• **PTH**   PTH is viewed as the primary regulator of osteoclastic activity; however, it stimulates bone deposition by osteoblasts as well. This paradox is based on the cellular localization of PTH receptors.

PTH receptors are not found on osteoclasts; PTH receptors are present on osteoblasts creating the primary link of bone deposition and resorption in the ARF cycle and establishing the basic concept of bone turnover.

The effects of PTH on bone formation and resorption differ depending on the concentration of PTH in the bloodstream. At low PTH levels bone formation by osteoblasts is stimulated, at higher levels osteoblasts are stimulated to release osteoclast-differentiation factors.

PTH stimulates differentiation of monocyte precursors to form osteoclasts. It also stimulates the formation of ruffled borders on osteoclasts.

The mechanism of osteoclastic regulation involves the binding of PTH to osteoblastic PTH receptors, resulting in the release of osteoclast differentiation factors. Elevated PTH levels result in increased bone resorption, leading to elevated calcium levels in the blood; excess PTH results in eroded bone and fibrosis of the resulting spaces (osteitis fibrosa). Targets for PTH include bone lining cells, osteoblasts, and osteoclasts (indirectly through osteoblasts). Bone lining cells are observed in the periosteum and endosteum and are capable of rapidly transporting Ca^{2+} from the bone fluid to the extracellular fluid (Fig. 14-4). The overall effect is an increase in bone resorption, depleting the Ca^{2+} reservoir of bone and resulting in elevated Ca^{2+} levels in the bloodstream. This regulation of Ca^{2+} levels can be accomplished without resetting the equilibrium point.

• CALCITONIN

Calcitonin is a potent, albeit transient, inhibitor of osteoclastic bone resorption. Calcitonin reduces blood calcium levels, but PTH is the primary regulator of calcium in the bloodstream.

Calcitonin acts to reduce bone resorption. It is important to note that calcitonin, like parathyroid hormone affects all osteogenic cells, not only the osteoclasts. There are calcitonin receptors on osteoclasts; however, calcitonin stimulates the differentiation of osteoblasts as well as inhibiting the action of osteoclasts. The net result of its action is a decrease of calcium levels in the bloodstream. Decreased osteoclastic activity results in inhibition of Ca^{2+} release into the bloodstream, and increased osteoblastic activity results in increased Ca^{2+} movement from the blood to be stored in the bone reservoir. While calcitonin opposes PTH, it is not an equal force and, unlike PTH, is not essential. In fact, reduction of PTH levels is more effective than calcitonin in reducing blood Ca^{2+} levels.

• **VITAMIN D (1,25-DIHYDROXYVITAMIN D$_3$)**   Vitamin D stimulates osteoblastic matrix and alkaline phosphatase synthesis, mineralization, and the production of bone-specific proteins. Vitamin D also stimulates osteoclastic differentiation. The scenario is similar to that for PTH: there are vitamin D receptors on osteoblasts but not on osteoclasts. Again, as in the case of PTH, osteoblasts function as the intermediary.

Vitamin D is unique in that it is produced in the skin by the action of sunlight (ultraviolet B radiation). In this newly synthesized form it is inert and must be converted by two successive hydroxylations in the liver and kidney to form dihydroxyvitamin D (1,25-dihydroxyvitamin D$_3$), which is biologically active.

The major target of vitamin D is actually not bone, but the small intestine where it increases the efficiency of calcium absorption.

Vitamin D and PTH feed back on one another. PTH is the primary regulator of dihydroxyvitamin D production by the kidney. Elevated 1,25-dihydroxyvitamin D results in increased intestinal transport of Ca^{2+} and, therefore, elevated blood Ca^{2+} levels. These in turn result in decreased PTH levels.

• **GROWTH FACTORS AND CYTOKINES**   Local factors (cytokines) such as interleukin 1 (IL-1), tumor necrosis factor (TNF), and transforming growth factor-beta (TGFβ) appear to be involved in the regulation of bone cells. IL-1 stimulates osteoclastic differentiation from monocytes as well as increasing the activity of already differentiated osteoclasts. IL-1 has been identified as an intermediate in the menopause-associated bone loss which occurs during osteoporosis. (Estrogen does not appear to have a direct effect on bone resorption.)

Growth factors are active in regulation of osteoblastic activity. Growth factors produced by bone or isolated from bone matrix include TGF-β, bone morphogenetic protein (BMP), and insulin-like growth factor-1 (IGF-1). TGFβ is present in bone in its inactive form, proTGFβ. Increased osteoclastic activity results in increased acid production, which stimulates TGFβ production from proTGFβ. This growth factor inhibits osteoclastic activity and increases osteoblastic activity. It provides a negative feedback link between osteoclastic and osteoblastic activity which might be useful in breaking the ARF cycle in *osteoporosis*. In that disease, there is a decrease in mineralized bone mass resulting from the failure of osteoblastic activity to keep up with osteoclastic activity. Osteoporosis affects both men and women with increasing age, but is most prevalent in women after menopause. BMP is another member of the TGF superfamily of growth factors and has been used to stimulate bone formation *in vitro* and in both experimental animals and humans *in vivo*.

Mechanisms of calcification

> Alkaline phosphatase formed by osteoblasts and hypertrophied chondrocytes releases PO_4^{2-} from organic phosphate compounds causing precipitation of $(Ca)_3 (PO_4)_2$. This product is in an amorphous form which is later modified to form the final crystalline product $Ca_{10}(PO_4)_6(OH)_2$ (hydroxyapatite), which makes up the mineralized component of bone.

Two mechanisms for bone mineralization have been described: use of matrix vesicles (developing bone) and mineralization in association with a collagen-noncollagenons protein complex (adult bone).

• **MATRIX VESICLES**   Calcified cartilage and woven bone mineralize by use of matrix vesicles which bud off from cells such as hypertrophied chondrocytes and osteoblasts. The matrix vesicles accumulate mineral and then release proteins that control the rate at which crystallization proceeds. Calcification is stimulated by several characteristics of the matrix vesicles: (1) high concentration of Ca^{2+}-binding acidic phospholipids, (2) presence of alkaline phosphatase on vesicle membranes which yield PO_4^{2-} for the formation of Ca_3PO_4, and (3) protection of the first nucleated crystals of mineral. Subsequently, formed hydroxyapatite crystals are exposed to the extracellular fluid. The vesicular crystals serve as seed crystals between collagen fibrils of the matrix.

> The matrix vesicles are necessary for the initiation of mineralization. The continuation of calcification is related to the removal of inhibitors of calcification such as a component of the cell membrane or a crystal inhibitor (e.g. pyrophosphate).

• **MINERALIZATION IN ADULT BONE**   In adult bone, matrix vesicles are rare. Lamellar bone contains extensive collagen fibrils associated with noncollagenous proteins such as proteoglycans and osteonectin. Mineralization develops in association with the collagen-noncollagenous protein complex. There appear to be three-dimensional holes formed in the heteropolymers of collagen and noncollagenous protein, which allow for entry of the inorganic ions. The periodicity of collagen appears to be important; however, collagen is a poor initiator of mineral deposition by itself.

Some physiologic factors affecting bone development

Parathyroid hormone is the primary regulator of blood calcium and has a major influence on bone. There are other hormones, vitamins, and growth factors which help regulate bone development.

• **VITAMIN D**   Deficiency of vitamin D results in impaired calcium absorption and disturbed phosphate metabolism. Ultimately this results in deficiency in cartilage calcification and bone matrix and eventually rickets.

• **VITAMIN A**   Vitamin A deficiency results in disturbed endochondral ossification and bone remodeling.

• **VITAMIN C**   Vitamin C deficiency (scurvy) results in decreased production of normal collagen fibers in the bone matrix.

• **GROWTH HORMONE** Deficiency in growth hormone results in dwarfism; excess results in gigantism or, in adults, acromegaly.

• **SEX AND THYROID HORMONES**   Sex hormones influence secondary ossification centers and the time of epiphyseal closure. Thyroid hormones act directly on osteoclasts altering cellular activity, but also affect bone and cartilage growth secondarily to their control of metabolic rate. Estrogen prevents excessive bone resorption, but IL-1 appears to be an intermediate in this process.

BONE REPAIR (FRACTURE HEALING)

In the formation of the fracture, bone is broken into two or more parts, each of which is called a bone fragment. The trauma tears soft tissues and blood vessels. Circulation ceases in Haversian vessels for some distance around the fracture line, resulting in death of osteocytes in the region. Some periosteal and marrow tissue also dies, although the better blood supply of these areas results in less extensive destruction than in the bone itself. The initial step in bone repair is *clot formation*. Proliferating fibroblasts, osteogenic cells, and budding capillaries (*granulation tissue*) invade the clot and help organize the clot into a *callus* (Fig. 14-9). There are various stages in callus formation (procallus, temporary callus, and bridge callus), although the only functionally important steps in callus formation are the *cartilaginous and bony callus*. The healing process closely resembles the original osteogenic process, and osteogenic cells will differentiate into osteoblasts or chondrocytes on the basis of the immediate microenvironment of the differentiating cells.

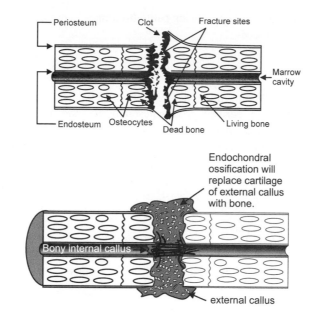

Figure 14-9 Bone repair.
(*A*) The response of bone immediately following a fracture. The first step in fracture healing is the formation of a clot which becomes organized to form an internal (bony) and external (cartilaginous) callus (*B*).

The *cartilaginous callus (external callus)* is formed around the bone fragments uniting the fragments. (*Remember*: blood supply in this area is poorest resulting in mesenchymal cell differentiation into chondroblasts and chondrocytes and the elaboration of cartilage matrix.) As blood supply (capillary growth) increases and enters the area, cartilage is replaced by bone. The process strongly resembles fetal endochondral ossification. Osteoblasts appear in the deep osteogenic layer of the periosteum and also the endosteum. Woven bone is laid down and begins to replace the cartilage. Remodeling to form lamellar bone occurs at a later time. ARF cycles occur in a similar fashion to bone development.

The *bony callus (internal callus)* is formed between the ends of the bone fragments and between the two marrow cavities where the blood supply is better than in the area of the external callus.

Bone grafts are commonly used to bridge gross fractures. This bone does not persist, and its periosteum provides only a minor contribution to repair. Osteogenic tissue of the host provides the major source of cells used in reparative processes. New bone spreads over the dead ends of the host bone and over the transplanted bone. Eventually, both the dead ends and the transplant are replaced by new bone. The transplant serves primarily as a temporary bridge and pathway for healing. Transplants including growth factors like BMP or a cocktail of growth factors appear to be a promising future treatment scheme for accelerating the fracture healing process.

Summary of bone cell interactions

Figure 14-10 depicts the interrelationships of bone cells. The osteogenic cell, or osteoprogenitor cell, is derived from mesenchyme and may give rise to osteoblasts and osteocytes in microenvironments where capillaries are present. This line of differentiation is also favored in the presence of the hormone calci-

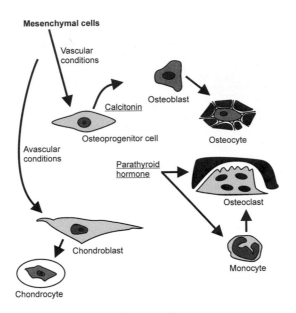

Figure 14-10 The interrelationships of bone cells.
Mesenchymal cells in the sclerotome region of the somite are pluripotential and can differentiate into cartilage or bone cells depending on environmental conditions. In the absence of a blood supply (avascular conditions), chondrocytes develop and form the extracellular matrix of a specific cartilage type. Under vascular conditions the mesenchymal cells differentiate into osteoprogenitor cells, which eventually form osteoblasts and osteocytes. Osteoclasts differentiate along a separate lineage from bone marrow–derived monocytes. The balance between osteoblastic activity and osteoclastic activity as well as the differentiation of the two bone lineages is related to hormonal levels. High levels of parathyroid hormone (PTH) are produced in response to low serum Ca^{2+} levels. PTH stimulates differentiation of osteoclasts and increases ruffled border formation on existing osteoclasts. Low PTH levels stimulate osteoblastic activity. High levels of calcitonin (produced by the C cells of the thyroid) also stimulate osteoblastic activity while inhibiting the differentiation of osteoclasts from monocytes. (Not drawn to scale or intended to describe the precise microscopic features of the cell types).

tonin (produced by the C cells of the thyroid gland). In the presence of parathyroid hormone, produced by the chief cells of the parathyroid, differentiation of osteoclasts from monocytes, increased extent of the ruffled border in osteoclasts, increased transport of Ca^{2+} to the extracellular fluid by bone lining cells, and osteocytic osteolysis (absorption of bone by osteocytes) occurs. Although it was once believed that osteoclasts developed from osteoprogenitor cells or from osteoblasts by cell fusion, there is now convincing evidence for the origin of osteoclasts from monocytes. Under avascular conditions mesenchymal cells or osteoprogenitor cells differentiate into chondrocytes and form a cartilaginous matrix.

MUSCLE

·

- **Skeletal Muscle**
- **Cardiac Muscle**
- **Smooth Muscle Cells**
- **Regeneration of Muscle Cells**
- **Nonmuscle Contractile Cells**

·　　·　　·　　·　　·　　·　　·　　·　　·　　·　　·　　·

Muscle is a specialized tissue in which the intracellular movement of thick filaments (myosin motors) along thin filaments (actin) (Chap. 8, "Cell Motility") is converted to extracellular movement or work. The three types of muscle tissue (skeletal, cardiac and smooth) share this basic characteristic, although finer points of function and structure may differ.

SKELETAL MUSCLE

Skeletal muscle consists of a network of individual muscle cells which join to form a functional unit.

Individual *myoblasts* join lengthwise during embryologic development to form a *myotube*, which matures into a cylindrical cell (*myocyte*) 50 to 60 μm in diameter with hundreds of nuclei. In mature skeletal muscle cells, the nuclei are displaced toward the periphery and lie just deep to the cell membrane (*sarcolemma*), whereas the *myofibrils* consisting of many sarcomeres linked in series (Chap. 8) occupy most of the rest of the cytoplasm (Fig. 15-1). Mitochondria, glycogen deposits, and lipid droplets are also abundant and provide the energy for contraction.

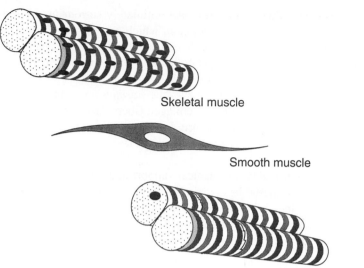

Skeletal muscle

Smooth muscle

Cardiac muscle

Figure 15-1 Classification of muscle types. (Not to scale.)
Skeletal muscle cells (*top*) are characterized by striations and peripherally located multiple nuclei. Striations are caused by the alignment of sarcomeres. Smooth muscle cells (*middle*) are smaller, somewhat spindle-shaped, and lack striations. Cardiac muscle cells (*bottom*) are characterized by striations, centrally located nuclei, and intercalated disks.

Skeletal muscle cells possess an elaborate tubular (t-tubule) system for spreading the depolarization of the cell membrane to the interior of the cell.

Nervous stimulation at a neuromuscular junction (see below) results in depolarization of the surface membrane which is carried to the interior of the myocyte by numerous invaginations of the cell membrane (*t tubules*). The t tubules reach deeply into the cell and make contact with paired elements of the *endoplasmic [sarcoplasmic (SR)] reticulum* (one t tubule plus two SR cisternae = *a triad*). In skeletal muscle, there are two triads per sarcomere, one associated with each A-I band interface. Depolarization of the SR cisternae by the t tubule results in the release of Ca^{2+} sequestered in the SR. As discussed in Chap. 8, calcium regulates the contraction of both striated and smooth muscle.

Connective tissue organizes the bundling of individual skeletal muscle cells into fascicles and the arrangement of many fascicles into a complete muscle.

Skeletal muscle fibers (cells) are each intimately surrounded by a delicate layer of connective tissue, the *endomysium*, which consists primarily of a basal (external) lamina secreted by the muscle cells, and reticular collagen fibers. This layer anchors the muscle fibers to each other and helps to distribute the mechanical force of contraction over the entire muscle.

A thicker layer of connective tissue, the *perimysium*, consisting of fibroblasts (plus other CT cells) and type I collagen fibers, envelops groups of skeletal muscle cells to form *fascicles*. In turn, fascicles are grouped together, and the entire muscle is surrounded by an even thicker layer of connective tissue, the *epimysium*.

In addition to providing mechanical support and distribution of contractile forces, the connective tissue layers (similar to those found in peripheral nerves, endo-, peri-, and epineurium), provide a conduit for blood and lymphatic vessels, as well as nerves, to reach the internal elements of the muscle.

Skeletal muscle fibers are classified by level of metabolism and speed of contraction.

When a cross section of skeletal muscle is stained for various oxidative enzymes and other molecules such as reduced nicotinamide-adenine dinucleotide (NADH) transferase, *myoglobin*, and ATPase, a mosaic pattern of staining results. Muscle fibers rich in the above substances appear dark (mainly due to myoglobin) and are termed *type I fibers*. They manufacture energy primarily through oxidative metabolism and produce slow but continuous contractions. Type IIA fibers stain for oxidative enzymes with intermediate intensity and produce ATP by both glycolysis and oxidative phosphorylation. They contract faster than type I fibers and are resistant to fatigue. Type IIB fibers are fast contracting but fatigue quickly. They are light-staining and produce ATP mainly by glycolysis. The proportion of type I to type II fibers in a given muscle in a specific individual is normally constant.

In skeletal muscle, each myocyte is innervated. The association between a peripheral motor neuron and a myocyte is termed the *neuromuscular junction (NMJ)*.

As a nerve fiber approaches a skeletal muscle cell, the axons lose their myelin sheaths and divide into fine terminal branches or telodendria. Each terminal branch distributes along a single myocyte and eventually comes to lie in a shallow groove (*primary cleft*) on the myocyte surface (Fig. 15-2). Each primary

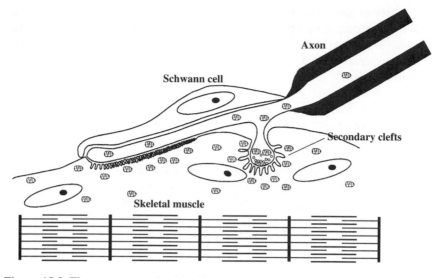

Figure 15-2 The neuromuscular junction.
Terminal processes (telodendria) of neurons occupy shallow depressions (primary clefts)
on the surface of a skeletal muscle cell. Secondary clefts increase the surface area of the
receptor portion of this modified synapse. Synaptic vesicles in the axon termini release
acetylcholine into the clefts when the neuron is stimulated. Muscle nuclei and mitochondria tend to accumulate in the region of the junction.

cleft gives rise to numerous *secondary clefts* which dramatically increase the surface area for reception of the neurotransmitter (*acetylcholine* in this case). The
axon terminal and the primary and secondary clefts, along with the surface of the
myocyte, constitute the neuromuscular junction (NMJ). The external lamina of
the muscle cell continues into the secondary clefts where it contains the enzyme
acetylcholinesterase to rapidly inactivate the neurotransmitter and prevent prolonged contraction of the muscle. A single axon and all the skeletal muscle cells
innervated by its branches constitute a *motor unit*. A motor unit may contain only
a few muscle cells (fine movement control) or over 100 (coarse movement control).

Skeletal muscle is also innervated by sensory nerves.

Sensory nerves relay information about a muscle's degree of stretch and tension to the CNS. The specialized sensory organs, or *muscle spindles*, consist of a
small group of thin muscle fibers (*intrafusal fibers*) separated from the regular
skeletal muscle fibers (*extrafusal fibers*) by a connective tissue capsule. Sensory
(afferent) fibers wrap around the intrafusal fibers and relay information about

their degree of stretch to the spinal cord where a reflex arc is established with small (γ) motor neurons which feed back to the intrafusal fibers to modify their tension. The sensory afferents also connect with large (α) motor neurons which regulate the tension in the entire muscle. Sensory receptors also transmit pain and touch to the CNS.

Skeletal muscles are connected to bone by tendons.

Tendons are composed of parallel arrays of collagen fibers oriented in the direction of tension (dense regular connective tissue). There are few fibroblasts and a scanty vascular supply. A fibrocollagenous covering of the tendon is continuous with the epimysium of the muscle. Bundles of collagen fibers are interdigitated with terminal infoldings of muscle fibers at the myotendinous junctions. At the other end of the tendon, the collagen fiber bundles insert into the periosteum of the bone.

CARDIAC MUSCLE

Cardiac muscle is similar to skeletal muscle in that both possess striations due to the alignment of myofibrils. In contrast to skeletal muscle, cardiac muscle consists of individual cells usually with only one centrally placed nucleus. Cardiac muscle possesses a t-tubule system that is less elaborate than that in skeletal muscle with the tubules interacting with only one SR cisterna (*diad*) in the region of the Z line at either end of each sarcomere.

Cardiac muscle cells are linked end-to-end by intercalated disks containing a variety of anchoring and communicating intercellular junctions.

Intercalated disks consist of macula adherentes (desmosomes), fascia adherentes, and gap junctions (Fig. 15-3). Desmosomes link the intermediate filaments (desmin) of adjoining cardiac muscle cells, whereas the fascial junctions anchor the actin microfilaments of the sarcomeres at either end of the cell. The gap junctions allow communication between adjoining cells and facilitate the conduction of an action impulse throughout the population of cardiac muscle cells.

Specialized cardiac muscle cells participate in the origin and conduction of cardiac contractile impulses.

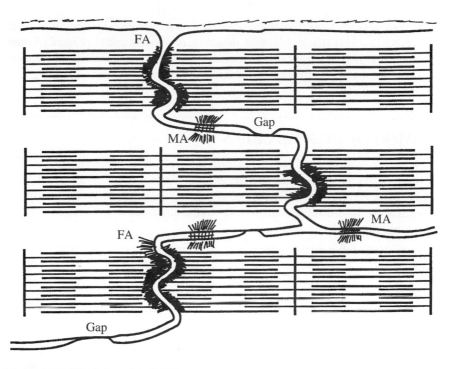

Figure 15-3 The intercalated disk.
Intercalated disks are found only between cardiac muscle cells. They consist of three types of intercellular junctions: fascia adherens (FA), macula adherens (desmosomes, MA), and gap junction.

Specialized cardiac muscle cells of the *sinoatrial (SA) node* have inherent rhythmicity and are the pacemakers of cardiac contraction. Their impulses are transmitted to other atrial cardiomyocytes through the gap junctions of the intercalated disks. Once the impulse reaches the *atrioventricular (AV) node*, it is conducted to the ventricular myocardial cells by other specialized cardiac muscle cells, the *Purkinje cells*. These cells are characterized by large size, few sarcomeres, abundant mitochondria, and glycogen deposits. After entering the interventricular septum as a large *bundle (of His)*, these fibers divide into right and left branches, decrease in size and ultimately blend imperceptively with ordinary ventricular cardiac muscle cells.

Some cardiac muscle cells have an endocrine function.

Cardiac muscle cells of the small chambers of the heart (atria) secrete *atrial natriuretic peptides (ANP)* which regulate fluid and electrolyte balance in the body, and relax vascular smooth muscle and thus reduce blood volume and pressure. ANP is stored as a prohormone in secretory vesicles found primarily in the perinuclear region of atrial cardiomyocytes. Release of ANP into the circulation is stimulated by atrial stretch and results in cleavage of the prohormone into a smaller circulating form.

SMOOTH MUSCLE CELLS

Smooth muscle cells are small, fusiform cells (Fig. 15-1) found in the walls of blood vessels and virtually all the internal organ systems. While sharing the basic contractile mechanism of skeletal and cardiac muscles, smooth muscle cells incorporate actin and myosin bundles into a meshwork arrangement.

Smooth muscle lacks the organized sarcomeres and resulting striations characteristics of striated (cardiac and skeletal) muscles.

Smooth muscle cells contain actin and myosin bundles arranged in a criss-crossing network throughout the cytoplasm except in the nuclear area. The actin filaments are anchored to each other and to the cell membrane by *dense bodies* which contain α-actinin and other proteins found in the Z lines of striated muscle. The dense bodies are connected to each other by the muscle-specific intermediate protein, desmin (Fig. 15-4). The dense bodies thus communicate the force of contraction to the cytoskeleton and the cell membrane.

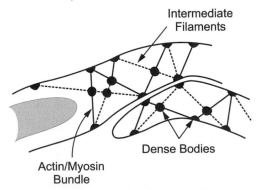

Figure 15-4 The contractile and structural apparatus of smooth muscle cells.
Actin-myosin bundles (solid lines) are attached to the cell membrane and each other by dense bodies. Intermediate filaments of desmin (dashed lines) stabilize the contractile apparatus.

Smooth muscle cells lack a t-tubule system and possess only a rudimentary sarcoplasmic reticulum.

Smooth muscle cells are, therefore, more dependent on extracellular calcium than on intracellular calcium stores for contraction. A system of cell surface invaginations has been observed which may shuttle extracellular Ca^{2+} to cisternae of the smooth ER. As described previously in Chap. 8, upon excitation of the smooth muscle cell, calcium ions bind with calmodulin to activate myosin light chain kinase, which subsequently phosphorylates the myosin light chain and initiates binding to actin.

Smooth muscle cells are sparsely innervated by both the sympathetic and parasympathetic components of the autonomic nervous system (ANS).

In contrast to that in skeletal muscle, contraction in smooth muscle may be either facilitated or inhibited by its innervation. In general, sympathetic innervation (adrenergic) and parasympathetic innervation (cholinergic) have opposite effects on smooth muscle. The adrenergic and muscarinic neurotransmitter receptors are linked to G proteins (Chap. 9), whereas the nicotinic cholinergic receptor is a cation channel. Various neuropeptide hormones and other agents also modulate contraction in smooth muscle cells. Visceral or "unitary" smooth muscle cells occur in sheets in the walls of the major organs. They generally receive a poor innervation but spread excitation through numerous gap junctions. The innervation tends to moderate intrinsic activity in these populations of smooth muscle cells. Multiunit smooth muscles, such as those found in the iris and the vas deferens, receive a much richer innervation, which provides precise control of contraction.

REGENERATION OF MUSCLE CELLS

Cardiac muscle cells have almost no regenerative capacity.

As a response to increased workload, the ventricles of the heart increase in wall thickness. This increase is due to proliferation of fibroblasts (*hyperplasia*) and resultant increase in the amount of connective tissue. In addition, the size of individual cardiac muscle cells increases (*hypertrophy*), but *cardiac muscle cells do not proliferate*. Similarly, cardiac muscle cells do not regenerate in response to ischemic injury (infarct) in the heart wall.

> Skeletal muscle is regenerated by the proliferation and differentiation of satellite cells.

Satellite cells are considered to be a population of inactive skeletal muscle *precursor cells* (myoblasts). They are normally found in small numbers lying between mature skeletal muscle cells and their external lamina. Minor injury or increased workload can activate these cells to proliferate and fuse with the existing syncytial myocytes. Enlargement (hypertrophy) of existing skeletal muscle also occurs after repetitive exercise.

> Smooth muscle cells possess a high capacity for proliferation and regeneration.

Both mature smooth muscle cells and *pericytes*, a pluripotential stem cell found in the vicinity of small blood vessels and capable of differentiating into myoblasts, proliferate in response to injury or increased vascular workload as occurs in hypertension. The increased mitotic activity (hyperplasia) adds new smooth muscle cells and results in hypertrophy of the vascular wall. In similar fashion, the smooth muscle cells of the uterine wall undergo extensive hyperplasia and hypertrophy during pregnancy.

NONMUSCLE CONTRACTILE CELLS

Cells of epithelial origin help glands expel their secretions into ducts.

Myoepithelial cells are spindle-shaped or stellate cells which possess actin and myosin filaments as well as tropomyosin. Their epithelial rather than mesodermal origin is confirmed by the presence of the intermediate filament protein, *keratin*, which is diagnostic for epithelial cells, rather than *desmin*, which is found in muscle cells. Myoepithelial cells are located between the basal lamina and the basal surface of acinar and ductal cells, particularly those in *mammary*, *salivary*, and *sweat glands*.

Other nonmuscle cells with contractile properties include the fibroblast-like cells of the perineurium and the intraglomerular mesangial (IGM) cells of the kidney. The myofibroblasts of the perineurium are thought to mediate retraction of the nerve fibers from a wound, whereas the IGM cells may regulate circulation through the capillaries of the renal glomerulus (Chap. 24).

NEURAL TISSUE

·

- **Excitability**
- **Conductivity**
- **Myelination**
- **Synapse**
- **Synaptic Vesicle Interactions With the Presynaptic Membrane**
- **Regeneration**

·　·　·　·　·　·　·　·　·　·　·　·

Nervous tissue is distributed throughout the body and functions as an integrative and communication network. Nervous tissue possesses two critical and highly developed characteristics, excitability and conductivity of impulses along its network. Neural tissue also has a limited ability for regeneration and repair. Nervous tissue is classified into peripheral and central nervous systems, which are discussed in Chap. 27.

> The neuron is the fundamental structural and functional unit of the nervous system and consists of a cell body (soma or perikaryon), dendrites, and an axon.

The basic structure of a neuron is shown in Fig. 16-1.

> The cell body of the neuron contains a single nucleus and all of the organelles found in most cells.

The cell body contains extensive protein synthetic machinery in the form of rough endoplasmic reticulum (RER). This RER may be so extensive that collectively it is visible in the light microscope as the Nissl substance. In addition, the

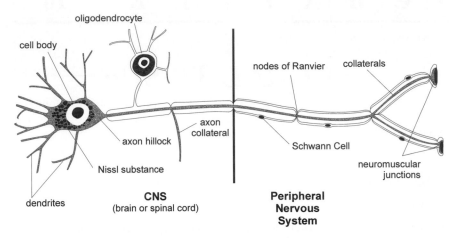

Figure 16-1 Motor neuron traversing central (CNS) and peripheral nervous systems to reach neuromuscular junctions.
The parts of the neuron are the dendrites, axon, and cell body (perikaryon). The Nissl substance is the granular material visible at the light microscopic level, corresponding to RER. It is absent from the axon hillock. Myelination occurs by oligodendrocytes in the CNS and Schwann cells in the peripheral nervous system.

neuronal cytoplasm contains elaborate Golgi apparatuses, mitochondria, microtubules, microfilaments, and all other organelles enveloped by a plasma membrane. Lipofucsin (pigment) granules accumulate in neuronal perikarya with age.

> The axon is a single process specialized for conduction of impulses to other neurons, whereas dendrites are single to multiple processes specialized for receipt of incoming impulses from receptors or other neurons.

Axons and dendrites have different morphologic and functional characteristics. These include the presence of ribosomes, mRNAs, and certain types of ion channels in the dendrites and cell body of the neuron, whereas there are axon-specific cell adhesion molecules and ion (Na^+) channels which are localized only in the axon. Differences between axons and dendrites are established through axon and dendrite-specific orientation of microtubules and distribution of microtubule-associated proteins.

Nissl substance is abundant in the dendrites as well as the cell body, but is absent from the axon and the transition region from the cell body to the axon, known as the *axon hillock*.

The long processes which traverse foot-long distances of the body require an extensive system of microtubules for the transport of molecules along the axon (axonal transport) or the dendrites (dendritic transport).

Axonal (axoplasmic) transport is a bidirectional transport system for molecules to be shipped between the axons and the cell body.

Shipping in the direction from the cell body toward the axon terminal is known as *anterograde transport*. Transport in the opposite direction from the axon terminal to the cell body is known as *retrograde transport*.

Anterograde transport resembles a three-speed transmission, with slow, intermediate, and rapid rates of transport. Enzymes, microfilaments, and microtubules are transported at the slowest rate, mitochondria and other organelles are shipped at an intermediate rate, and vesicular transport of molecules is the most rapid form of axonal transport. Rapid anterograde transport and retrograde transport use microtubule motors.

Slow axonal transport occurs at a rate of 1 to 6 mm per day. Enzymes and cytoskeletal components are transported by slow transport. Flow occurs only in the direction away from the cell body toward the axon terminal. There are actually two rates of transport: slow component a (SCa, which includes transport of preassembled microtubules and neurofilaments) and slow component b (SCb, which includes transport of enzymes, actin, and clathrin). Intermediate transport occurs at a rate of 50 to 100 mm per day (mitochondria and other membrane-bound organelles). Fast transport occurs at 200 to 400 mm per day (synaptic vesicles and neurotransmitters). The most rapid form of anterograde axonal transport uses the microtubule motor kinesin to transport neurotransmitter-containing vesicles to the axon terminal. Retrograde transport of endocytosed materials and recycled proteins from the axon terminal move toward the cell body at a rate of 100 to 300 mm per day, using dynein as a motor.

Dendritic transport differs from axonal transport in regard to the types of microtubule-associated proteins found in dendrites versus axons and the orientation of microtubules. Dendritic transport occurs at a rate of about 3 mm/hour.

Neurons are classified on the basis of their morphologic appearance as pseudounipolar, bipolar, or multipolar.

Pseudounipolar neurons have one process which bifurcates into a central and peripheral process. They are found in the ganglia alongside the spinal cord (dorsal root ganglia). Bipolar neurons have two processes, one at each end of the spindle-shaped neuron, and are found in association with special senses (e.g. retinal ganglion cells). Multipolar neurons have multiple processes and are the most common morphologic type.

Neurons are also classified based on their function: motor (efferent), sensory (afferent), or interneurons.

Motor neurons carry impulses to muscles or other effectors such as glands. Sensory neurons carry impulses from receptors to the central nervous system. Interneurons connect sensory and motor neurons and may be used to send information to higher or lower levels.

Not all cells in nervous tissue are neurons. Glial cells are the supportive, nutritive cells of nervous tissue. The types of glia include astrocytes, oligodendrocytes, Schwann cells, ependymal cells, and microglia.

The glia differ in embryologic origin, location, and function as shown in the table.

ORIGIN, LOCATION, AND FUNCTION OF GLIA

GLIAL TYPE	EMBRYOLOGIC ORIGIN	LOCATION	FUNCTION
Astrocytes	Neuroepithelium	Central nervous system (CNS)	K^+ sink, structural support, metabolism of neuronal byproducts, initiate blood-brain-barrier formation, secrete trophic factors for neurons
Oligodendrocytes	Neuroepithelium	CNS	Forms myelin in the CNS
Microglia	Bone marrow	CNS	Phagocytosis, presentation of antigen
Schwann cells	Neural crest	Peripheral nervous system (PNS)	Forms myelin in the PNS
Ependymal cells	Neuroepithelium	Lining ventricular system of CNS	Transport functions
Satellite cells	Neural crest	Moons around cell body (planet) in ganglia	Insulators

EXCITABILITY

Excitability is a property of all protoplasm, but is heightened in neurons because of the low resting membrane potential (unequal charge across the membrane—from inside to outside).

CONDUCTIVITY

The plasma membrane of the neuron maintains a constant membrane potential of -90 mV. This membrane potential is maintained by ion pumps distributed throughout the membrane. The Na^+,K^+-ATPase removes sodium from the inside of the neuron and exchanges it for potassium. Thus K^+ is in high concentration inside the neuron, whereas Na^+ is high in concentration outside the neuron at resting potential. Gated channels are of two types: voltage-gated channels, which respond to depolarization, and ligand-gated channels, which open or close in response to binding of neurotransmitters. When depolarization begins, the voltage-gated Na^+ channels open, leading to an explosive influx of Na^+ resulting in increased membrane potential and the production of an action potential which is transmitted down the axon. The action potential is an all-or-none phenomenon, building and building until the threshold is reached. It occurs with constant duration and amplitude in a particular axon. The K^+ channels also open in response to depolarization, but their function is to move the membrane to the hyperpolarized state. The movement of K^+ combined with the closing of the sodium channels returns the membrane to the resting potential. The speed of conduction is dependent on the electrical characteristics of the axon which are modified by myelination.

MYELINATION

Myelination is a wrapping of membranes which insulates the axon and allows for more rapid conduction down the axon.

Myelination of the axon provides insulation which prevents leakage of current and improves axonal electrical characteristics, allowing for increased speed of conduction down the axon. Myelination occurs during development in synchrony with the maturation of neural tracts. Therefore, those tracts which are incomplete at birth are myelinated postnatally (after birth).

Myelination occurs by a similar process in the central and peripheral nervous systems, but the cells responsible for myelination are different.

In the peripheral nervous system the cells which form myelin are the *Schwann cells*. A single Schwann cell becomes married to a 1- to 2-mm segment of axon. Their relationship is intimate and monogamous as the Schwann cell rolls around the axonal segment producing a series of flattened membranes with high lipid content. The squeezing of Schwann cell cytoplasm from the stacks results in alternating major dense lines separating amorphous protein material when a myelinated fiber is viewed by electron microscopy. In the central nervous system, there is a polygamous relationship between the *oligodendrocyte* and the axons it myelinates: a single oligodendrocyte may be responsible for the myelination of numerous axons. In the central nervous system (CNS) one oligodendrocyte process wraps one segment of an axon, whereas in the peripheral nervous system (PNS) one segment is wrapped by a single Schwann cell.

There are also differences in the oligodendrocyte-myelin and Schwann cell–myelin relationship resulting in minor structural differences (e.g., the presence of the displaced Schwann cell cytoplasm within the myelin, forming the Schmidt-Lanterman clefts in the peripheral nervous system).

The difference between a myelinated and unmyelinated fiber is the ratio of oligodendrocyte processes to nerve segments or Schwann cells to nerve segments in the CNS and PNS, respectively.

Unmyelinated fibers are associated with more than one oligodendrocyte process or Schwann cell. This is the critical difference between myelinated and unmyelinated fibers. The result is that the Schwann cell or oligodendrocyte process in unable to wrap around the fiber segment, and myelination does not occur.

Preparation of tissue for microscopy generally involves dehydration with alcohol, which also solubilizes and removes lipid from the myelin sheath having an apparent clear space between the axon and an intimate investing layer of connective tissue, the *endoneurium*.

The regions between the myelin segments are known as the *nodes of Ranvier*. These regions, like the myelin itself, are specialized for rapid conduction of impulses.

In a myelinated fiber the action potential travels from node to node (saltatory conduction).

The result is rapid depolarization with leakage of sodium limited primarily to the nodal region because this is the region with the highest concentration of sodium-gated channels. These channels are anchored via ankyrin to the cytoskeleton (see Chap. 3). This system is very efficient because leakage of Na^+ only in the nodal area reduces the energy required to pump Na^+ back out of the axon.

SYNAPSE

Communication between neurons occurs in regions known as *synapses*. The majority of synapses use chemical mediators (neurotransmitters). There is no contact between the pre- and postsynaptic fibers; communication is by release of neurotransmitters from the presynaptic endings, which has an effect on the polarization of the postsynaptic membrane.

Synapses are primarily the interrelationship of an axon with the dendrites (axodendritic) of another neuron, although axons frequently synapse with the soma (axosomatic) of an adjacent neuron. Axoaxonic synapses between two axons also occur, although less frequently than the other types.

The propagation of an action potential results in the opening of *voltage gated Ca^{2+} channels* in the synaptic terminal. The influx of Ca^{2+} stimulates the release of neurotransmitters from vesicles on the presynaptic side of the membrane.

Synapses are of two main types: inhibitory and excitatory. In an inhibitory synapse the release of neurotransmitters results in hyperpolarization of the postsynaptic membrane, whereas in a stimulatory synapse the result of neurotransmitter release is the depolarization of the postsynaptic membrane and the conduction of the nerve impulse to the adjacent neuron.

The neuromuscular junction is discussed in Chap. 15, "Muscle," and is an example of a modified synapse.

SYNAPTIC VESICLE INTERACTIONS WITH THE PRESYNAPTIC MEMBRANE

Synaptic vesicles contain neurotransmitter and are transported by a rapid transport mechanism to the axon terminal. How are the vesicular-presynaptic membrane interactions regulated?

Migration toward the presynaptic membrane, as well as docking and fusion of the synaptic vesicles with the presynaptic membrane, are mediated by specific proteins called SNAPs. These important proteins are found in the cytosol of the terminal while SNAP receptors (integral membrane proteins) are found in the presynaptic and synaptic vesicular membranes.

The SNAPs [soluble NSF (N-ethylmaleimide-sensitive fusion protein) attachment proteins] are intermediates which bind the synaptic vesicles to the presynaptic membrane. This was provided as an example of vesicular interactions with the membrane during secretion (Chap. 6, "Intracellular Trafficking").

Once released into the synaptic cleft, neurotransmitters bind to receptors on the postsynaptic terminal, resulting in opening of ligand-gated Na^+ channels. Propagation of local depolarization from the postsynaptic terminal toward the soma is accomplished by *voltage-gated channels*. Once reaching the soma, depolarizing and hyperpolarizing events are summed spatially and temporally. If summation results in net depolarization of a critical level, an *action potential* will be initiated.

REGENERATION

Wallerian degeneration follows injury to a peripheral nerve. Following injury the nucleus within the perikaryon moves peripherally, and Nissl is depleted, a process called *chromatolyis*.

The response of the axon distal and proximal to the injury differs dramatically.

Distal to the injury the nerve fiber and its associated myelin sheath degenerate and macrophages enter the area to phagocytose debris. Proximal to the injury there is repair as Schwann cells proliferate and form a tube. Where sprouting and growing axons traverse these Schwann cell cords, there can be successful regeneration and reinnervation of muscle. Axonal growth under these circumstances may occur at speeds up to 3 mm per day. Where axons are not successful in reestablishing connections, a neuroma forms and the muscle fiber is not reinnervated.

BLOOD AND BONE MARROW

·

- **Erythrocytes (Red Blood Cells)**
- **Leukocytes (White Blood Cells)**
- **Nomenclature**
- **Blood Clotting (Coagulation)**
- **Intrinsic Clotting Pathway**
- **Extrinsic Clotting Pathway**
- **Plasminogen Activator and Plasmin (Clot Dissolution)**
- **Bone Marrow: Blood Cell Development**

· · · · · · · · · · · ·

Blood is the river of life of the mammalian body. It is considered as a tissue and is classified as a specialized connective tissue because of the presence of cells in an extracellular matrix (fluid), known as the plasma. Blood carries cells and a variety of proteins through a closed circulatory system. People who weigh 150 lbs have approximately 5 to 6 L of blood in their circulatory systems.

Blood consists of cells and plasma.

The cells of the blood carry out diverse functions from oxygen transport to phagocytosis. Some function in the blood, whereas others enter tissues where they differentiate into other cell types or directly carry out specific functions. The plasma contains proteins, inorganic salts, and other molecules passing through (amino acids, vitamins, hormones, antibodies, blood clotting proteins, and a variety of other molecules). Calcium is an important ingredient of plasma at a relatively constant concentration of 1 mg per 10 mL of blood.

> The cells of the blood include red blood cells (erythrocytes), white blood cells (leukocytes), and platelets.

The cells of the blood (erythrocytes, leukocytes, and platelets) are sometimes called the *formed elements*, but the platelets are actually fragments of cells. Even the most common cell of the body, the erythrocyte, lacks a nucleus (*enucleated*) and might be considered "less than a cell." The red blood cells (RBCs) contain hemoglobin, an oxygen-carrying pigment. The white blood cells function in a diversity of protective mechanisms as the body is being attacked by bacteria, viruses, or allergens. White blood cells function in phagocytosis, directly attacking foreign agents, as well as secretion of substances which serve a role in defense. *Platelets* are circulating, cellular fragments which release blood clotting factors and other growth factors. They are an essential part of the clotting mechanism.

> Serum is plasma without the coagulation factors.

Blood clots when removed from blood vessels. If clotted blood is centrifuged, the precipitate is the cells, and the clear, yellowish fluid supernatant is the serum. Serum contains many of the molecules found in the plasma except for the clotting factors, which are part of the clot and are excluded from the serum.

> The most common and useful way to separate and analyze the blood and its components is to add an anticoagulant (e.g., heparin) to blood removed from the body. The treated blood can then be centrifuged. The result is three layers: (1) supernatant (plasma), (2) buffy coat (leukocytes), and (3) precipitate (sedimented red blood cells).

Plasma contains a variety of essential proteins: fibrinogen, albumin, globulins (α, β, and γ globulins) as well as lipids. Fibrinogen is a globulin and a target for the proteolytic enzyme, thrombin, which converts fibrinogen to fibrin, the scaffolding of a blood clot. Albumin is the primary protein constituent of the blood and is important for maintenance of osmotic pressure. α globulins, β globulins, and γ globulins (also known as *immunoglobulins* or antibodies), play an essential role in the body's immune mechanisms. The *formative elements* ("cells") make up 45 percent of the blood. The volume of packed erythrocytes per unit volume of blood (sedimented RBCs) is defined as the *hematocrit*. The hematocrit averages 47 percent in normal men and 42 percent in normal women.

ERYTHROCYTES (RED BLOOD CELLS)

Erythrocytes are enucleated cells which carry oxygen and carbon dioxide and are morphologically adapted for gas exchange between the blood and the tissues and organs of the body.

Red blood cells are the most common cell in the blood (9.5×10^6 in males and 4.5×10^6 in females). Under the influence of the hormone *erythropoietin*, produced by the kidney, the number of erythrocytes increases following exposure to an oxygen-poor environment. Mature erythrocytes have a diameter of about 7.5 μm and therefore are useful for determination of size in microscopy. The biconcave shape of the erythrocyte allows for flexibility in passing through the smaller capillaries where gas exchange is optimal, maximizes surface area for gas exchange, and maintains the low viscosity of the blood. The flexibility of the membrane is established through membrane fluidity provided by cholesterol. The biconcave shape is maintained through the activity of the protein spectrin, which is bound to the integral membrane proteins by ankyrin. Hemoglobin in the RBC can combine with oxygen to form oxyhemoglobin or carbon dioxide to form carbaminohemoglobin. In contrast, the toxic gas carbon monoxide binds to hemoglobin forming carboxyhemoglobin, often leading to irreversible reduction in transport capacity.

LEUKOCYTES (WHITE BLOOD CELLS)

White blood cells have little known function in the blood, but play important roles in the tissues.

It is not a coincidence that white blood cells are some of the same cells which are found in the connective tissue. These cells leave the blood and are targeted to specific regions of a variety of organs. The roles of white blood cells include phagocytosis and involvement in *cellular immunity* (cell-antigen interactions) and *humoral immunity* (antibody-antigen interactions), which are discussed in Chap. 19, "The Immune System."

The normal range for numbers of leukocytes is 5000 to 9000 white blood cells/per milliliter of blood. A decrease in the number of leukocytes below 5000 mL is called *leukopenia*, and an increase above this normal range is known as *leukocytosis*.

Leukocytes are classified into two classes, agranular and granular. The agranular leukocytes include lymphocytes and monocytes. The granular leukocytes include neutrophils, basophils, and eosinophils.

This classification ignores the fact that agranular cells may contain a few granules in the cytoplasm. The characteristics of leukocytes are summarized in the table.

LEUKOCYTE CHARACTERISTICS

CELL	SIZE (μM)	CHARACTERISTICS	FUNCTION(S)
Small lymphocyte	6–8	Nucleus occupies most of cell	Role in chronic inflammation
Medium lymphocyte	10–12	A rim of cytoplasm	Role in chronic inflammation
Large lymphocyte	Up to 18	Clumped chromatin; more cytoplasm than small and medium lymphocytes	Precursors of T and B lymphocytes
B-Lymphocyte	6–18	Depends on size	Precursor of plasma cells, source of antibodies (immunoglobulins) used in humoral immunity
T-Lymphocyte	6–18	Depends on size	Precursor of T cells which modulate other immune cells (through interleukin release), kill virus-infected and tumor cells
Monocyte	9–12	Eccentrically placed, ovoid, kidney, or horseshoe-shaped nucleus, a few cytoplasmic granules	Precursor of macrophages and osteoclasts
Neutrophil (polymorphonuclear leukocyte, PMN)	7–9	3–5-lobed nucleus, tiny ("specific") granules containing antibacterial proteins and larger ("azurophilic") granules which are lysosomes	First line of defense, phagocytosis, particularly of bacteria
Eosinophil	9–10	Bilobed nucleus, eosinophilic granules (containing major basic protein and an array of lysosomal enzymes)	Respond in allergic diseases (phagocytosis of antigen-antibody complex) and parasitic infections phagocytose parasites); granules contain histaminase
Basophil	7–9	Bilobed nucleus, obscured by basophilic granules containing heparin, histamine, and eosinophilic chemotactic factor	Contains heparin (anticoagulant) and (increases vascular permeability)

The morphology of the white blood cells (leukocytes) is shown in Fig. 17-1.

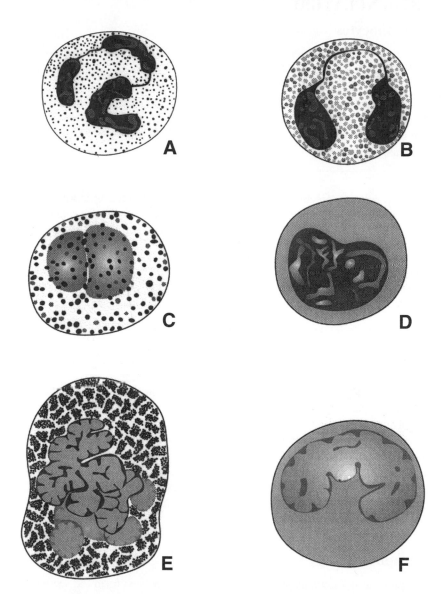

Figure 17-1 The most common blood-related cells.
All these cells are found in the blood except for *E* (the megakaryocyte). *A* is an illustration
of a neutrophil with a typical 3–5 lobed nucleus, *B* a basophil with granules covering the
nucleus, and *C* an eosinophil with a bilobed nucleus. *D* is an illustration of a monocyte
(ropy chromatin; horseshoe-shaped nucleus), *E* a megakaryocyte, and *F* a large lymphocyte.

NOMENCLATURE

Lymphocytes

Lymphocytes are characterized by the presence of a round nucleus with increasing cytoplasm and loss of roundness as one moves from small to medium to large lymphocytes. The designation of lymphocytes as small, medium, and large is based on size and the amount of cytoplasm visible. Large lymphocytes are the source of either B or T lymphocytes. These cells are often called simply B and T cells, derived from the bone marrow and thymus, respectively.

The B cells are precursors of plasma cells which produce antibodies. The T cells are involved in modulation of other immune cells through the production of interleukins (T-helper cells), death of virus-infected cells (cytotoxic T cells), and death of tumor cells (natural killer cells).

Monocytes

Monocytes are the precursors of a lineage known as the *monocyte-phagocyte system.*

Specific phagocytic cells, called the microglia in the brain, Kupffer cells in the liver, and Langerhans cells in the dermis of the skin, as well as macrophages throughout the body, are derived from monocytes at some time during the life of the animal. Monocytes have little, if any, phagocytic function in the blood. They differentiate into macrophages in a variety of tissues and organs. Macrophages are capable of phagocytosis of dead and dying cells and debris as well as other immune functions. In bone, monocytes differentiate into osteoclasts, cells responsible for bone resorption through localized acidification of the environment.

Neutrophils

Neutrophils, also known as polymorphonuclear leukocytes, form the second line of defense for the body.

Neutrophils phagocytose bacteria.

One of their roles is the phagocytosis of bacteria, although their effectiveness varies with the type of bacterial infection. Neutrophils kill bacteria by formation of free radicals (superoxide, O_2^-), as well as the release of lysozyme and lactoferrin, which destroy bacterial cell walls and interfere with the normal nutrition of bacteria, respectively.

Eosinophils

Eosinophils, as the name implies, contain granules which stain brightly with eosin or other acidophilic dyes.

Eosinophils modulate the immune response in allergic reactions by releasing histaminase, which breaks down the histamine produced by basophils and mast cells. Eosinophils like to chew on (phagocytose) antigen-antibody complexes and parasites.

Eosinophils respond during hay fever and asthma in which basophils and mast cells are also active.

Basophils

Basophils are very similar to mast cells seen in the connective tissue and mucosa. They are rare in blood smears (0.5 percent). Basophils contain heparin and histamine, but are increased in very few pathologic conditions, such as smallpox and chicken pox.

Heparin prevents blood coagulation, and histamine causes vasodilation, resulting in a "red flare," as well as increased capillary permeability. Histamine causes rapid swelling and hives, which occur soon after exposure to an allergen.

Platelets

Platelets, like erythrocytes, are not really cells. They are derived from megakaryocytes and are cell fragments. Platelets are about 2 μm in diameter and are found in a concentration of 200,000 to 400,000/per microliter of blood. Megakaryocytes are large cells in the bone marrow, which are the source of platelets. The megakaryocytes never enter the bloodstream.

The platelets are the orchestral conductor of hemostasis (prevention of blood loss). They enhance aggregation by release of factors, and they promote clot formation, retraction, and dissolution.

Platelets aggregate to repair damage (see table) to the endothelium (epithelial lining of blood vessels), a process known as formation of the *platelet plug*. Small damage results in repair (closure) by the plug. Where damage is more extensive, a *clot* forms. Adhesion of platelets involves the integrins (Chap. 8). Factors in the damaged endothelium and/or soluble factors stimulate β_3-integrins in the platelet plasma membrane altering adhesive properties of the platelets to (1) the damaged area stimulating plug formation and (2) fibrinogen, which plays a role in clot formation.

FACTORS WHICH REGULATE PLATELET AGGREGATION

FACTOR	SOURCE	FUNCTION	G PROTEIN
Thromboxane	Platelets	Platelet aggregation ↑	Gq
Prostacyclin	Endothelial cells	Platelet aggregation ↓	Gs

Thromboxane and prostacyclin function through different G proteins and have opposing effects on platelet aggregation.

Platelets take up serotonin and release it into the bloodstream, resulting in vascular contraction. Platelets also release *platelet-derived growth factor (PDGF)*, which causes proliferation of smooth muscle cells and plays a major role in wound healing. Retraction and dissolution involve reshaping and reorganization of the clot through the shortening of fibrin strands bringing together the edges of the blood vessel wall and sealing the injury site. This clot resolution prevents obstruction of normal blood flow and is regulated by the *plasmin* system.

BLOOD CLOTTING (COAGULATION)

The blood clotting process consists of five distinct steps: (1) prothrombin activation; (2) conversion of prothrombin to thrombin; (3) thrombin conversion of fibrinogen to fibrin, the mesh work in which platelets, blood cells, and plasma become entrapped to form the actual clot; (4) reshaping of the clot by polymerization of fibrin, and (5) dissolution of fibrin clots through the activation of the plasminogen activator system and the action of plasmin.

Prothrombin activation occurs by both intrinsic (through the circulation) and extrinsic (tissue interaction) mechanisms (see Fig. 17-2). The extrinsic and intrinsic pathways are separate but overlapping and lead to a common pathway. The second step is formation of thrombin with activated factors X and V (X_a and V_a) in the presence of Ca^{2+} and phospholipid. Thrombin is a proteolytic enzyme which cleaves specific peptide bonds in fibrinogen to form fibrin monomers. The third step requires the proteolytic activity of thrombin on fibrinogen to form fibrin strands (monomers). The fourth step is the reshaping of the fibrin clot through polymerization of the monomers to form a fibrin polymer. The last step, clot dissolution, is regulated by the plasmin system.

> The prothrombin activation pathway is formed by a series of zymogen or proenzyme activations (procoagulants).

Each step results in the activation of an inactive enzyme. The resulting, activated clotting factor catalyzes the activation of the next factor in the cascade. Each step also requires an enzyme, a substrate on which the enzyme works, and a cofactor such as Ca^{2+} or other membrane factors.

> There are two pathways which lead to thrombin activation. Although described as separate mechanisms, there is overlap and interaction between the systems.

Some courses require that students learn the details of the clotting pathway so the key coagulation factors and their functions are illustrated in Fig. 17-2. That figure also shows the major steps involved in prothrombin activation, as well as thrombin and fibrin formation.

INTRINSIC PATHWAY

The intrinsic pathway utilizes the circulation and is initiated by contact activation factors.

The intrinsic pathway begins with activation of factor XII. Kininogen and kallikrein are two contact activation factors (proteins) that have effects on the

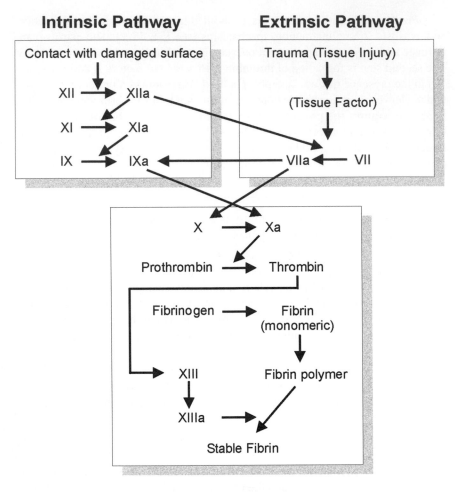

Common Pathway

Figure 17-2 Extrinsic and intrinsic coagulation pathways.
The X between intrinsic and extrinsic pathways denotes the convergence of those two
pathways into the common pathway leading to fibrin formation.

extrinsic pathway as well as other related regulatory systems such as bradykinin, which causes vasodilation. There are other examples of interactions between the extrinsic and intrinsic pathways. For example, XII_a catalyzes activation of factor VII $\rightarrow$ VII_a (along with tissue thromboplastin and Ca^{2+}). Activated factor XII (XII_a) catalyzes conversion of XI $\rightarrow$ XI_a, which in turn converts IX $\rightarrow$ IX_a. Factor IX_a catalyzes the activation of factor X, the first step in the common pathway.

EXTRINSIC PATHWAY

The extrinsic pathway is initiated by the entry of tissue thromboplastin into the circulation.

The extrinsic pathway involves the formation of tissue factor and entry of thromboplastin (phospholipoproteins and organelle membranes from cells in the disrupted tissue) into the circulating blood. Factor VII binds to platelet phospholipid and is activated to form VII_a. The complex of VII_a, tissue thromboplastin, and Ca^{2+} activates factor X in the common final pathway leading to thrombin. There is a feedback loop in that factor X_a interacts with factor V, Ca^{2+}, and phospholipid to produce thromboplastin. Thromboplastin also is activated by other factors in the intrinsic pathway.

Figure 17-2 illustrates the extrinsic and intrinsic coagulation pathways. The primary intersection point of the two pathways is factor X and its activation factor (X_a). That is why, in the figure, the enlarged X marks the spot.

PLASMINOGEN ACTIVATOR AND PLASMIN (CLOT DISSOLUTION)

Dissolution of fibrin clots begins soon after they are formed. *Plasmin* digests fibrin by hydrolysis to form fibrin fragments. *Plasminogen* is the inactive form of plasmin which circulates in the plasma. Plasminogen is activated by the action of proteolytic enzymes (plasminogen activators), which stimulate the conversion of plasminogen to plasmin. These molecules include *tissue plasminogen activator (tPA)* and *urokinase (uPA)*. Plasminogen activator inhibitor prevents the formation of plasmin. This is another "yin and yang" system of balance between the activators and the activator inhibitors. The role of these factors is discussed again in Chap. 18 in relation to the function of endothelial cells.

BONE MARROW: BLOOD CELL DEVELOPMENT

The bone marrow laboratory is the dreaded laboratory of the histology course because of the fine differences between the different developmental stages and the horrendous classification schemes for developing erythrocytes and leukocytes. The key features of bone marrow cell type identification are presented in this section, which focuses on the bone marrow as an example of renewal by *pluripotential stem cells*. A similar pattern is illustrated by the small intestinal crypt progenitor cell (Chap. 21).

The bone marrow is the site of hematopoiesis and is responsible for the continuous replacement of the blood cells.

At birth the marrow of all bones is *red bone marrow* and the source of blood cells. With time there is a restriction of hematopoietic sites to the sternum and iliac crest with the marrow of long bones changing to fat (*yellow marrow*). Yellow bone marrow can be transformed back into hematopoietic (red) bone marrow under conditions of hemorrhage or hypoxia.

The marrow consists of sinusoidal capillaries and hematopoietic cells, in a cordlike arrangement, supported by a connective tissue stroma.

The stroma contains the usual extracellular matrix (ECM) molecules with specialized ECM molecules like hemonectin, which bind the hematopoietic cells in cords.

The ECM functions in anchoring and release of hematopoietic cells, binding of growth factors, and homing of progenitor/stem cells.

Smaller, mature cells from each of the lineages pass between the endothelial cells into the sinusoids. The release process is regulated by hormones such as glucocorticoids, C3 component of complement, and toxins derived from bacteria. The release and transport into the sinusoid differs for the blood cell lineages. For example, maturing leukocytes are highly mobile, whereas transport of red blood cells may be due to a pressure or other type of gradient. The squeezing through the endothelium and stroma is also believed to be responsible for the enucleation process of RBCs in the bone marrow. Megakaryocytes are very large cells visible at low magnification on the light microscope. These cells cannot travel across the sinusoidal epithelium and end up fragmenting into platelets.

> In the bone marrow, there is a pluripotential stem cell which is capable of self-renewal or differentiation into a committed progenitor (stem) cell. The committed progenitors are called *colony forming cells (CFCs)* and indicate a degree of determination.

The committed progenitors are dedicated to a specific lineage. For example, a CFC-T is committed to the thymus and the T-cell lineage, whereas CFC-B cell differentiates within the bone marrow to form B cells. Similarly, there are CFC-eosin (eosinophils), CFC-bas (basophils), CFC-GM (neutrophils and monocytes), CFC-MEG (megakaryocytes), and CFC-E (erythrocytes) cells. The regulation of CFCs is under the control of colony stimulating factors (see table).

COLONY STIMULATING FACTORS

COLONY STIMULATING FACTOR	SOURCE	TARGET
Erythropoietin	Kidney	CFC-E
Interleukin-3	T lymphocytes, keratinocytes	Pluripotential and committed stem cells
Granulocyte/macrophage	T lymphocytes, endothelial cells, fibroblasts	CFC-GM
Granulocyte	Macrophages	CFC-G
Macrophage	Macrophages, endothelial cells	CFC-M
Stem cell (steel) factor	Bone marrow stromal cells	CFCs

In addition to the colony stimulating factors, other growth factors (often called cytokines) such as interleukin-1 (IL-1) and interleukin-6 (IL-6) affect more than one CFC. Transforming growth factor beta (TGFβ) inhibits cell proliferation in the bone marrow, downregulates colony stimulating factors and growth factor receptors, and interferes with other growth factors.

The terminology for bone marrow cells and the hemopoietic steps in differentiation of blood cells are downright confusing. This is especially true for erythropoiesis, in which the bone marrow progenitors have three sets of names. *The general concept of maturation and cell renewal in the bone marrow involves loss of pluripotentiality, decreased mitotic activity, and morphologic characteristics of decrease in cell and nuclear size and increased condensation of chromatin.* The classification schemes are listed in the tables for maturation of erythrocytes and granulocytes.

ERYTHROPOIESIS

PREFERRED NOMENCLATURE	PREVIOUSLY USED TERMS—1	PREVIOUSLY USED TERMS—2
Proerythroblast	Rubriblast	Rubriblast
Early normoblast	Prorubricyte	Basophilic erythroblast
Intermediate normoblast	Rubricyte	Polychromatophilic erythroblast
Late normoblast	Metarubricyte	Orthochromatophilic erythroblast
Reticulocyte*	Reticulocyte	Reticulocyte
Erythrocyte*	Erythrocyte	Erythrocyte

*These cells are normally found in the circulation.

GRANULOPOIESIS

Myeloblast
Promyelocyte (last stage common to all three granular leukocyte lineages)
Myelocyte (neutrophilic, basophilic, or eosinophilic)
Metamyelocyte (neutrophilic, basophilic, or eosinophilic)
Band (neutrophilic, basophilic, or eosinophilic)*
Mature neutrophil, basophil, or eosinophil*

*These cells are normally found in the circulation.

Identification of cells in the bone marrow is viewed as one of the most horrific experiences of the histology laboratory. Figures 17-3 and 17-4 serve to simplify the identification of cell types by specific criteria. It must be remembered that the bone marrow contains cells undergoing continual differentiation with changes in both the nucleus and cytoplasm occurring simultaneously. *It is important to start with ideal examples of specific cell types using these characteristics and not to try to identify every cell in the field.*

Figure 17-3 illustrates the pattern of granulopoiesis for eosinophils. The lineage shown in Fig. 17-3 illustrates eosinophilic maturation within the bone mar-

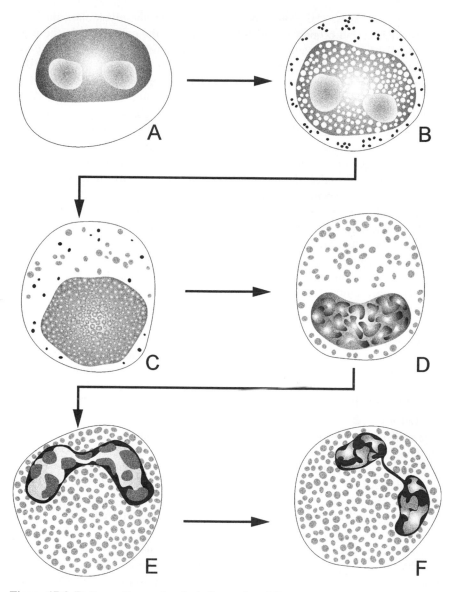

Figure 17-3 Pattern of granulopoiesis for eosinophils.

row. The first stage in granulopoiesis is the *myeloblast* (Fig. 17-3*A*), a large cell with prominent, light-staining nucleoli with only a little cytoplasm, generally without granules. The *promyelocyte* (Fig. 17-3*B*) is the next cell in the lineage. The first two stages (myeloblast and promyelocyte) are identical for all three granulocyte lineages. The promyelocyte is larger than the myeloblast, nucleoli are less visible, and primary granules are present in the cytoplasm. In the next stage, specific granules (neutrophilic for the maturing neutrophil, eosinophilic for the maturing eosinophil, and basophilic for the maturing basophil) are seen in the cytoplasm along with a flattening of the nucleus on one side. The eosinophilic *myelocyte* (Fig. 17-3C) differentiates into the eosinophilic *metamyelocyte* (Fig. 17-3D) when the nucleus begins to indent. Further invagination results in a large indentation and stretching of the nucleus into a band shape, resulting in a band eosinophil and ultimately a mature eosinophil. The metamyelocyte to *band stage* progression has similar nuclear shape changes in each of the lineages. The *mature eosinophil* is bilobed rather than having the three to five lobes seen in the *neutrophil* (polymorphonuclear leukocyte).

Figure 17-4A shows a *proerythroblast* with a round nucleus containing delicate chromatin, and a reddish stain and two to four dark-stained nucleoli. The cytoplasm has a thin rim of dark, royal blue cytoplasm in a stained slide. The next stage is the early *normoblast* (B) with a round nucleus and a coarsely clumped "checkerboard" pattern; the nucleoli are not visible and the cytoplasm is intensely blue, although it loses its royal blue appearance. The next stage is the *intermediate normoblast* (C), which has a round nucleus, with clumped chromatin. Nucleoli are not present, and the cytoplasm is blue-gray or gray-green reflecting a mixture of ribosomes (the protein-producing machinery) and hemoglobin. The next stage is the *late normoblast* (D), which contains a round pyknotic (very small, dense) nucleus, and the cytoplasm is essentially the same color as the mature erythrocytes, which are seen in the background of a standard preparation (Wright's or Giemsa stain) of a bone marrow smear. The *retriculocyte* (E) is an intermediate step in which the nucleus has been extruded in the bone marrow and the cytoplasm contains a netlike structure with an almost granular appearance, which is lost during the final maturation to an erythrocyte. The reticulocyte count in the peripheral blood may be used to determine erythrocyte production. An elevated reticulocyte count (reticulocytosis) is an indication of a shortened erythrocyte life span. The *mature erythrocyte* (F) has no nucleus and has a buff to reddish cytoplasm. RBCs have a biconcave shape, although this may not be observed in blood and bone marrow smears.

The bone marrow uses a process seen in continuously renewing populations and during development to correct for errors in renewal, produce a balance between cell production and death, and counteract excessive production of cells. This process is known as programmed cell death, or *apoptosis*. In the bone marrow, the arrangement of progenitor cells is critical. For example, maturing erythrocyte progenitors are grouped together surrounded by macrophages which have a dual function, production of growth factors and phagocytosis of old erythrocytes.

Bone marrow transplantation (BMT) is used in treatment of genetic diseases, aplastic anemia (loss of cell renewal capability of the bone marrow), and

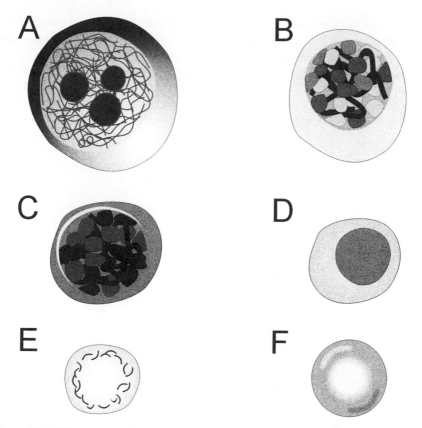

Figure 17-4 The pattern of erythropoiesis.

malignant diseases such as leukemias. BMT is being used more extensively in combination with high-dose chemotherapy in treatment of patients with high-risk tumors. The bone marrow may be taken from the same patient before treatment (*autologous* transplant). Other terms often used are *syngeneic* to classify a transplant in which the donor and recipient are identical (for example, identical twins) and *allogeneic* to classify a transplant in which donor and recipient are of different genetic origins. BMT utilizes an intravenous injection of bone marrow stem cells. These cells home to the bone marrow by specific signaling and recognition between cell surface molecules on the stem cells and specific receptors on the membrane of bone marrow sinusoidal endothelial cells (similar to selectin recognition discussed in Chap. 18). Once in the bone marrow there is a further interaction between the stem cells, stromal cells, and ECM molecules resulting in binding of the stem cells in place. The hematopoietic stem cells must home to the correct location so that they interact with appropriate growth factors and ECM molecules. Growth factor treatment has been used in a cocktail arrangement to improve homing and self-renewal of the hematopoietic stem cells.

HEART AND CARDIOVASCULAR SYSTEM

·

- **The Heart**
- **Blood Vessels**
- **The Tunica Intima**
- **The Tunica Media**
- **The Tunica Adventitia**

· · · · · · · · · · · ·

The heart pumps blood through a quasiclosed system of blood vessels consisting of arteries, capillaries and sinuses, and veins. However, the cardiovascular system is not merely a pump and a set of connected hoses. The walls of the system are not inert but secrete and respond to numerous vasoregulatory factors. Various components of the system including cardiac muscle cells, smooth muscle cells, and vascular endothelial cells secrete substances into the blood which, among other functions, alter blood pressure and regulate clotting.

THE HEART

The heart is enclosed in a double-walled membrane, the pericardium.

The *pericardium* is best conceptualized as a deflated balloon or a bean-bag chair with the heart indenting one side so deeply that it presses against the other side, reducing the airspace between the layers of the balloon to a minimum. The side of the balloon in intimate contact with the heart consists of a single layer of flattened epithelial (*mesothelial*) cells, the *visceral pericardium*. This layer also makes up the outermost portion of the epicardium (see below). The visceral pericardium is reflected around the root of the heart where the major blood vessels enter and leave to become the *parietal pericardium*. The parietal pericardium is thickened by an outer layer of tough, fibrous connective tissue (CT). Between the two pericardial layers is a potential space, the pericardial cavity or sac, occupied normally by a thin layer of lubricating fluid.

The wall of the heart is composed of three layers: epicardium, myocardium, and endocardium.

The *epicardium* is the outer covering of the heart. It consists of an outer single layer of squamous mesothelial cells (*visceral pericardium*) and an underlying layer of connective tissue (Fig. 18-1). That layer contains the major blood vessels supplying the heart (coronary arteries and cardiac veins/coronary sinus), nerves and ganglia, lymphatic vessels, and varying deposits of adipose tissue (depending to some degree on how conscientious a patient has been about diet and exercise regimen).

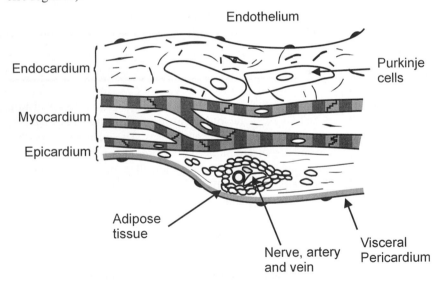

Figure 18-1 Schematic drawing of the heart wall.
The outer portion of the wall of the heart (epicardium) consists of a single layer of squamous mesothelial cells and underlying connective tissue containing adipose (fat) cells, nerves, arteries, and veins. The middle layer (myocardium) consists primarily of cardiac muscle cells, fibroblasts, and collagen. Only a few layers of myocardium are shown in this drawing. The endocardium consists of a layer of endothelial cells lining the chambers of the atria and ventricles, and underlying connective tissue which may contain components of the cardiac conduction system (Purkinje cells and the bundle of His).

The *myocardium* consists of thick layers of cardiac muscle cells (see Chap. 14) in various orientations (Fig. 18-1). Fibroblasts and collagen fibers are interspersed among the cardiac muscle cells. They provide structural support and distribute the force of contraction throughout the myocardium. Although the cardiac muscle cells are not connected to a bony support, they are connected to the dense connective tissue forming the valve rings and *fibrous trigones* (cardiac skeleton) near the base of the heart. This provides a substrate for contraction. The myocardium is highly vascular, with numerous capillaries interspersed among the cardiac muscle cells. Nerves, however, do not penetrate deeply into the myocardium.

The *endocardium* consists of a single layer of squamous endothelial cells continuous with the endothelial lining of the blood vessels entering or leaving the heart. Deep to this layer are layers of reticular and coarse connective tissue (subendocardial), occasional smooth muscle cells, components of the cardiac conduction system, and numerous small blood vessels.

Several types of modified cardiac muscle cells form the conducting system of the heart.

All cardiac muscle cells possess *intrinsic rhythmicity*. However, specialized cardiac muscle cells of the *sinoatrial (SA) node* are the fastest and control the beat rate of the others (*cardiac pacemaker*). The SA node is located in the wall of the right atrium near the opening of the superior vena cava. Nodal cells are smaller than atrial cardiomyocytes, lack intercalated discs and contain fewer myofibrils (Fig. 18-2). The cells of the SA node are circumferentially disposed around the

Figure 18-2 Specific types of cardiac muscle cells.

Ventricular myocytes are characterized by large, cylindrical fibers with elaborate intercalated discs. Atrial myocytes tend to be smaller, more ovoid, and have less intricate intercalated discs. Nodal cells are small and generally lack intercalated discs. Purkinje cells are large, have only scattered myofilaments, and possess abundant glycogen deposits.

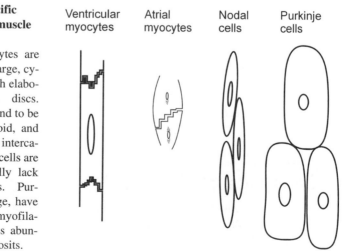

Ventricular myocytes   Atrial myocytes   Nodal cells   Purkinje cells

nodal artery and enclosed within a dense fibrocollagenous matrix. The contractile impulse originating in the SA node travels via gap junctions through other atrial muscle cells or through specific but histologically indistinguishable tracts of atrial myocytes to reach the *atrioventricular (AV) node* located adjacent to the right atrioventricular opening and the interatrial septum. AV nodal cells are histologically similar to SA cells but connect via gap junctions with a larger set of modified cardiac muscle cells, the *Purkinje fibers* (Fig. 18-2). The Purkinje fibers penetrate the membranous part of the interventricular septum as the *bundle of His* and divide into two main branches (bundles) in the subendocardial connective tissue of the muscular part of the septum, one going to each ventricle. Purkinje cells are wider than ventricular cardiomyocytes, are deficient in myofibrils, and contain abundant deposits of glycogen.

Atrial myocytes of the adult mammal synthesize and secrete a peptide with natriuretic, diuretic, and vasorelaxant properties.

Atrial natriuretic peptide (ANP) is synthesized as a preprohormone which, after removal of the signal sequence (see Chap. 5), is stored as a prohormone in perinuclear secretory vesicles. When stimulated by increased atrial stretch, the secretory vesicles are transported to the cell membrane and the prohormone is cleaved to the circulating form during the process of secretion. ANP increases sodium and water output from the kidneys, inhibits secretion of other hormones which increase body sodium and water content (aldosterone, renin, vasopressin), and acts as a vasodilator of most vascular beds. The family of natriuretic peptides now also includes brain natriuretic peptide (BNP) and C-type natriuretic peptide (CNP).

BLOOD VESSELS

Blood vessels (see classification table) are composed of three layers, the relative size of each layer reflecting the function of the vessel.

THE TUNICA INTIMA

The *tunica intima* is the innermost layer and lines the lumen (Fig. 18-3). It is present in all blood vessels.

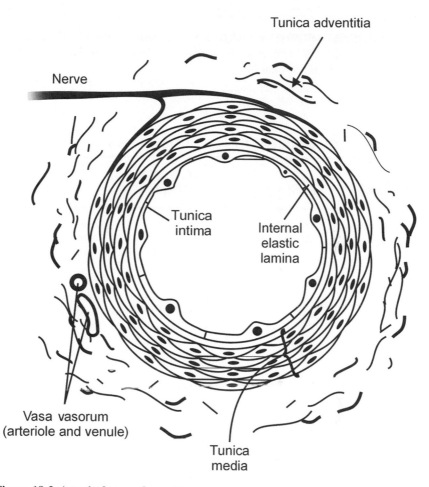

Figure 18-3 A typical muscular artery.
The tunica (t.) intima consists of a single layer of squamous endothelial cells (simple squamous epithelium) lining the lumen of the vessel. A layer of elastin (internal elastic lamina) separates the t. intima from the smooth muscle cells of the tunica media. A less constant layer of elastin fibers (external elastic lamina) may separate the t. media from the connective tissue forming the tunica adventitia. Nerves and small blood vessels (vasa vasorum) are found in the t. adventitia.

The tunica intima consists of a simple squamous epithelium (a single layer of flattened endothelial cells) and underlying subendothelial loose connective tissue including collagen fibers and elastin. In arterioles and muscular arteries, a well-defined layer of elastin, the *internal elastic lamina*, separates the tunica intima from the muscular tunica media.

The endothelial cells form a selective barrier between the blood and the interstitial fluid. The selectivity of the barrier depends on the particular requirements of the surrounding tissue.

CLASSIFICATION OF BLOOD VESSELS

TYPE	FUNCTION	CHARACTER OF WALL	NAME
Elastic artery	Conductance (high pressure, low volume)	Sheets of elastin between smooth muscle cells in tunica (t.) media	Aorta, pulmonary artery
Muscular artery	Distribution to organs and muscles	Smooth muscle cells predominate in t. media	Radial artery, popliteal artery
Arteriole	Resistance, distribution within organs	Little t. adventitia; 1–5 layers of smooth muscle cells in t. media	None
Metarteriole	Regulation of capillary flow	Discontinuous tunica media	None
Capillary	Exchange	No t. adventitia or t. media; pericytes may be present	None
Postcapillary venule	Capacitance (high volume, low pressure)	T. adventitia present but only scattered smooth muscle cells in t. media	None
Small to medium veins	Capacitance	Valves present	Cephalic vein, saphenous vein
Large veins	Capacitance	Prominent t. adventitia with longitudinal smooth muscle	Inferior vena cava

In most blood vessels, the endothelial cells form a continuous layer joined by tight junctions and gap junctions to prevent the interchange of substances across the vascular wall. Only water and small molecules can leak across the intercellular junctions (paracellular pathway).

The endothelial lining of capillaries may be continuous, fenestrated, or discontinuous (sinusoidal).

In *continuous capillaries*, the endothelial cells are joined by tight junctions which permit the passage of water and small solutes only. Small molecules may also diffuse directly across endothelial cells, whereas larger molecules may be shuttled in either direction by small pinocytotic vesicles. These types of capillaries are found in exocrine glands, muscle, and connective tissue. A continuous endothelium with only sparse pinocytotic vesicles and very tight intercellular junctions constitutes the *blood-brain barrier* of the nervous system.

Fenestrated endothelial cells are characteristic of capillaries found in endocrine glands, the kidney, and portions of the GI tract. These cells contain windows (fenestrae; visualize donut holes or the portholes in a cruise ship) which may be spanned by a diaphragm thinner than the plasma membrane. The diaphragms are *polyanionic*, consisting primarily of proteoglycans, particularly heparan sulfate, and thus present a charge barrier. Large anionic proteins such as insulin, albumin, transferrin, and low-density lipoprotein (LDL) are transported either through neutrally charged channels or plasmalemmal vesicles. The fenestrae permit the rapid passage of water and small solutes, but are too small to be penetrated by cells.

In capillaries lined with a *discontinuous* endothelium, the endothelial cells are not joined tightly and may contain many large fenestrae. In addition, the basal lamina is also interrupted. The interruptions between endothelial cells are large enough for cells to squeeze through. Discontinuous capillaries, or *sinusoids*, are typically found in blood-forming tissues including the liver and bone marrow, and the spleen, where all types of blood cells may enter or leave the circulation.

For their small size, endothelial cells are remarkable synthetic machines, synthesizing a variety of substances which regulate blood clotting and vascular blood flow (see table).

Endothelial cells contain the usual cytoplasmic organelles including rough endoplasmic reticulum (RER) and small perinuclear Golgi apparati. Endothelial cells of blood vessels larger than capillaries are also characterized by the presence of *Weibel-Palade bodies*, which are storage vesicles for *von Willebrand factor* (clotting factor VIII; see Chap. 17). Although many of the substances produced by endothelial cells are secreted constitutively, von Willebrand factor is released

PRODUCTS SYNTHESIZED BY ENDOTHELIAL CELLS*

CLASS		SPECIFIC FUNCTION
Anticoagulants		
	Prostacyclin (PGI$_2$) and nitric oxide (NO)	Inhibit adhesion of activated platelets to uninjured endothelium
	Thrombomodulin	Membrane-associated protein which converts thrombin to an anticoagulant
	Proteins C and S	Activated by thrombin-thrombomodulin to cleave clotting factors Va and VIIIa
	Heparin-like molecules	Stimulate inactivation of thrombin and other clotting factors by antithrombin III
	Tissue plasminogen activator (tPA)	Fibrinolysis
Procoagulants		
	von Willebrand factor (vWF, factor VIII)	Facilitates platelet adhesion to collagen
	Tissue factor	Activates extrinsic clotting pathway
	Plasminogen activator inhibitor (PAI)	Inhibits fibrinolysis
Vasoactive substances		
	NO	Vasodilator
	Prostacyclin	Vasodilator
	Endothelin	Vasoconstrictor
	Angiotensin converting enzyme (ACE)	Membrane-associated enzyme: converts angiotensin I to the vasoconstrictor, angiotensin II
Mediators of immune responses		
	Major histocompatibility class (MHC) II	Immunologic recognition when stimulated by cytokines to function in antigen presentation
	Adhesion molecules	Mediate binding (pavementing)
	Interleukins (IL-1, IL-6, and IL-8)	Mediation of smooth muscle responses

*Clotting factors are discussed in detail in Chapter 17 and immunity mediators are covered in Chapter 19.

via the regulated secretory pathway when stimulated by endothelial injury. A deficiency in this factor is one of many causes of hemophilia.

Another important group of factors noted in the table are *heparin* and the heparin-like molecules. These substances stimulate antithrombin III to inactivate thrombin and thus inhibit blood clotting. This has a practical use in hematology, since blood samples are usually collected and stored in tubes coated with heparin to prevent clotting (see Chap. 17).

Endothelial cells have regenerative capacity and form the foundation of new blood vessel formation and revascularization following wounding.

Although they have a relatively long lifespan (months to years) endothelial cells continue to divide slowly during adulthood. The first blood vessels in the embryo are formed entirely of intimal endothelial cells, the connective tissue and muscular components of the walls being added as required. Wounds to the endothelial lining are repaired by the proliferation and migration of neighboring endothelial cells. When new blood vessels are required, they sprout from existing small vessels by the extension of endothelial processes under the direction of extracellular matrix components and growth factors released by surrounding tissues. Vacuoles form in contiguous endothelial cells which then coalesce to form a hollow tube surrounded by endothelium. Thus all blood vessels begin developmentally as capillaries.

Endothelial cells play a major role in the immune response by regulating the passage of white blood cells from the blood into the tissue.

Normally, the tight junctions between endothelial cells prevent the escape (*extravasation*) of cells from the blood into underlying tissues. However, following damage to tissue, factors are released which lead to local loosening of endothelial cell junctions and the expression of a glycoprotein (*p-selectin*) on the lumenal surface of the endothelial cells. p-Selectin is a member of the selectin family of glycoproteins and contains a *lectin-binding domain* which binds specific carbohydrates (*oligosaccharides*) on the surface of white blood cells such as neutrophils and lymphocytes (Fig. 18-4). The weak binding of selectin signals the cortical actin of the WBCs to undergo a sol-to-gel state change, stiffening the cell and allowing it to roll along the endothelial cell surface until cells make stronger attachments via integrins (see Chap. 8). The WBCs then release proteolytic

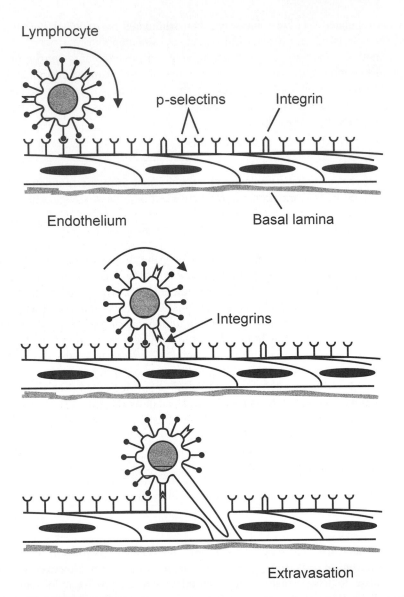

Figure 18-4 Extravasation.
Local damage induces the appearance of p-selectin on the lumenal surface of capillary endothelial cells. p-selectin binds to oligosaccharides on the lymphocyte surface (top panel). Binding the p-selectin initiates stiffening of the lymphocyte membrane and permits rolling of the lymphocyte along the surface until a stronger bond is made via integrins (middle panel). The integrin bond induces release of enzymes which further weaken the endothelial cell junctions and digest the underlying basal lamina, allowing the lymphocyte to squeeze between the cells by a process called diapedesis resulting in extravasation (bottom panel).

enzymes which digest the endothelial cell basal lamina and allow the WBCs to squeeze through the loosened junctions in the endothelium (*diapedesis*).

THE TUNICA MEDIA

The tunica media consists primarily of smooth muscle cells and varying amounts of collagen and elastin.

The smooth muscle cells of the tunica media (Fig. 18-3) secrete their own collagen and elastin since fibroblasts are not present. *Type III collagen* is found in the intercellular connective tissue whereas *types IV and V* are associated with the external laminae of the smooth muscle cells. *Elastin* is found in sheets of varying size between the layers of smooth muscle cells and in the internal and external elastic laminae bordering the tunica (t.) media. Smooth muscle cells are linked with each other by gap junctions and occasional desmosomes.

Capillaries possess a t. media consisting entirely of sparse *pericytes* loosely associated with the endothelial cells (see Chap. 14). Pericytes have some contractile function and are capable of differentiating into smooth muscle cells upon stimulation (e.g., by increased blood pressure). *Arterioles* possess one to three (or four or five in other textbooks) layers of smooth muscle cells in circumferential arrangement. Larger *muscular arteries* may contain dozens of layers of smooth muscle cells separated by incomplete sheets of elastin. In the large *elastic arteries*, elastin forms complete lamellae between layers of smooth muscle cells, which remain in contact with each other via processes extending through discontinuities in the elastic lamellae. The high elastin content of these vessels allows them, through elastic recoil, to dampen the pulsatile nature of the blood leaving the heart during systole.

Smooth muscle cells retain proliferative capacity and may undergo both hyperplasia and hypertrophy in response to chemical mediators as well as mechanical factors. Increase in size and number of smooth muscle cells occurs during physiologic adaptation (e.g., in the uterus during pregnancy) and pathologic damage (e.g., arterioles during hypertension).

Specialized smooth muscle cells function as an endocrine organ.

Specialized smooth muscle cells of the *afferent arterioles* of the renal glomerulus secrete *renin*, an important enzyme in the cascade which ultimately results in formation of *angiotensin II*, a potent vasoconstrictor. This is discussed in more detail in Chap. 24, "The Urinary System."

THE TUNICA ADVENTITIA

The tunica adventitia consists of fibroblasts, type I collagen and other CT matrix components, and adipose tissue.

The *tunica adventitia* (Fig. 18-3) contains numerous nerves which may penetrate the outermost layers of the tunica media, as well as blood vessels (*vasa vasorum*) which provide nutrients and waste removal for the outer layers of the tunica media in large vessels where these needs cannot be met by diffusion to and from the vessel lumen. The vasa vasorum are analogous to the coronary arteries of the heart which supply blood to the myocardium.

IMMUNE SYSTEM (LYMPHOID TISSUE AND THE FUNCTION OF IMMUNOLOGICALLY RESPONSIVE CELLS)

·

· · · · · · · · · · · ·

This chapter provides a fundamental histologic and cell biologic introduction to the relationships between lymphoid organs and tissues, cellular immunology, and inflammation. There is an excellent text by John Clancy, Jr, *Basic Concepts in Immunology* (McGraw-Hill, New York, 1998), that is dedicated to a more detailed description of the workings of the immune system.

The immune system provides defense against environmental pathogens. Immune cells are organized within the framework of lymphoid tissue. Included within the immune cell repertoire are cells specialized for phagocytosis, presentation of foreign molecules, killing, and production of defensive and recruitment molecules.

> The two main forms of immune defense are innate and acquired immunity. Innate immunity lacks immune specificity and memory, hallmarks of acquired immunity.

Some lymphoid tissue is responsible for production of cells for innate immunity, the initial hand-to-hand combat with foreign invaders. An innate immune response is described as *inflammation*. Many of the cell types involved in inflammation are described in Chap. 13, "Connective Tissue." These cells include macrophages, neutrophils, and eosinophils, as well as mast cells and basophils. Other lymphoid tissues are the source of cells for the acquired immune defenses initiated when invaders overcome innate checkpoints in the peripheral tissues. Lymphocytes are specifically involved in the acquired immune defense mechanisms.

> Acquired immunity develops in response to antigens, molecules which provoke an immune response (immunogenic).

Acquired immunity is more powerful than innate immunity, but takes longer to develop and has special initiation requirements. The immunogenicity of the foreign material (antigenicity) provides the specificity inherent in acquired immunity. Acquired immunity is also dependent and restricted by specific cell surface molecules encoded in one gene locus called *major histocompatibility complex (MHC)* discussed later in this chapter.

LYMPHOID TISSUE

> Lymphoid tissues provide structure for the immune system and a reservoir of cells that combat foreign invaders.

> Lymphoid tissue consists of a unique connective tissue framework formed by reticular fibers (composed of type III collagen) and connective tissue cells.

The cell biology of type III collagen is described in Chap. 13, "Connective Tissue." The immune cells are arranged within this framework. Lymphoid tissue appears in the body as a gradient from diffuse lymphoid tissue to aggregated lymphoid tissue (lymphoid nodules) to lymphoid organs. Diffuse lymphoid tissue and aggregates are transitory arrangements of cells that arrive in the connective tissue underlying the epithelium of many systems. They are particularly numerous in systems vulnerable to foreign invaders (gastrointestinal, skin, and respiratory systems).

> Lymphoid organs are categorized as primary or secondary based on involvement in lymphocyte education or in producing acquired immune responses.

The thymus and bone marrow function as primary lymphoid organs. Secondary lymphoid organs include the lymph nodes, spleen, and tonsils.

The lymph nodes are interspersed in the path of the lymphatic vessels. These vessels form as blind-ended tubes in the connective tissues and coalesce to form thin-walled lymphatic vessels. The lymphatics are more permeable than vascular capillaries (see Chap. 18), resulting in facilitation of movement of large molecules and cells from the connective tissue into the lymphatic circulation.

Many of the cells described in this chapter circulate between the connective tissue and the blood. Those cells include lymphocytes that aggregate in secondary lymphoid tissue to form a structural, spherical arrangement known as a *lymphoid nodule*, or *follicle*. When the nodule has a lighter-staining central portion (*germinal center*), it is called a *secondary lymphoid follicle*. In the absence of germinal centers the follicles are called *primary nodules*. The germinal center is an area of lymphocyte proliferation and activation in response to antigen. Lymphocytes may aggregate in tissues to form unencapsulated lymphoid nodules. Tonsils are partially encapsulated by an epithelium while lymph nodes and spleen are encapsulated by connective tissue. The type of encapsulation, one of the critical criteria for diagnostic identification of the lymphoid organs, is discussed in Chap. 28. Lymphoid tissues also contain a network of macrophages that are specialized to cooperate with lymphocytes in generating acquired immunity.

CELLS OF THE IMMUNE SYSTEM

All lymphocytes are born in the bone marrow. There are two types of lymphocytes: T lymphocytes that receive their primary education in the thymus and B lymphocytes that receive their primary education in the bone marrow.

The two primary lymphoid organs or schools for education of lymphocytes are the thymus for T lymphocytes (T cells) and the bone marrow for B lymphocytes (B cells). During development, naive T-lymphocyte stem cells (pre-T cells) travel from bone marrow to thymus where they receive their education. They develop the ability to recognize a wide range of environmental antigens. Antigen recognition is dependent on expression of antigen receptors on the T cell surface.

T and B lymphocytes are educated to recognize antigen during their development (maturation).

In the course of development, T lymphocytes acquire surface molecules that promote cell-cell interaction during generation of immune responses. At the same time, T lymphocytes undergo a selection process to eliminate cells that recognize and respond to self. When their education is complete, T lymphocytes acquire a receptor-dependent "exit visa" to leave the thymus and seed T-dependent regions of secondary lymphoid organs and tissues throughout the body. B lymphocytes are educated to recognize and respond to a wide range of environmental antigens while they are developing in the bone marrow. When B lymphocytes complete their primary education, they leave the bone marrow and undergo a similar trek to B-dependent areas of the secondary lymphoid organs. B- and T-lymphocyte travel between lymphoid compartments is dependent on specific recognition and recruitment signals similar to those for the seeding of bone marrow cells in a bone marrow transplant (see Chap. 17). The cell biologic mechanisms involved in recognition, recruitment, and entry into tissues are discussed in Chap. 18, specifically in the section dealing with the vasculature.

T lymphocytes are involved in acquired cell-mediated immune responses, whereas B lymphocytes are involved in acquired humoral (antibody-related) immune responses.

As with all generalizations, there are exceptions. The exception in this case is that T lymphocytes are required for generation of B-lymphocyte–dependent immune responses. This is a specific subset of T lymphocytes that provides help for the humoral response (helper T lymphocytes). *Cell-mediated responses* include processes such as viral immunity, whereas *humoral responses* include processes such as toxin neutralization.

B and T lymphocytes are involved in only acquired immunity. Other immune cells are involved in both innate and acquired immunity. Macrophages are involved in innate immunity as phagocytic scavenger cells. They also are involved in innate immunity as the source for a wide range of mediator molecules. Macrophages are involved in acquired humoral and cell-mediated immunity as antigen-presenting cells. Macrophages also are involved in acquired cell-mediated and humoral immunity as effector cells. They cooperate with B and T lymphocytes to carry out the effector phases of acquired immunity. Like macrophages, other bone-marrow–derived connective tissue cells play critical roles in the innate and acquired immune responses. Neutrophils (polymorphonuclear leukocytes, PMNs, or "polys") are the first cells to enter a site of inflammation. They phagocytose bacteria and release reactive oxygen species (superoxide and hydroxyl radicals). Their corpses form a major component of pus. Antibodies (discussed in the next section) are produced by B lymphocytes and cooperate with neutrophils to increase the efficiency of phagocytosis. Eosinophils are involved in phagocytosis during allergic reactions and parasitic infections. Mast cells contain histamine and heparin mediators that play a key role in immune responses, particularly those associated with allergy (immediate hypersensitivity). Basophils are similar to mast cells, but are found in the blood.

Antibodies

B lymphocytes are the parents of the antibody-producing cell lineage. Antigens are inducers of an immune response; they induce the generation of antibodies, hence the name *antigen* derived from the words *anti*body *gen*eration, or generator. The cells of the immune system respond to antigen by producing antibodies that subsequently bind specifically with the antigen.

Each antibody (see Fig. 19-1) has a highly variable portion for *recognition of antigen (Fab fragment)* and a "business" end (*Fc*, effector-binding portion) responsible for *binding to cells and a component of key serum proteins (complement)*. The immunoglobulins are all composed of *light and heavy chains*. Each immunoglobulin class has a class-specific heavy chain.

There are five main classes of antibodies, which differ in structure and function:

1. *IgG* is the principal immunoglobulin in the blood (serum) and is responsible for the vast majority of antibody function.

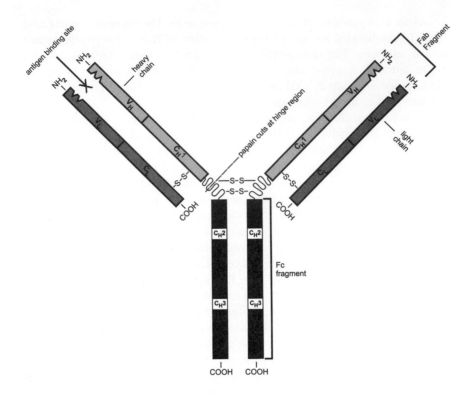

Structure of Antibody (IgG)

Figure 19-1 The general antibody structure.
The light and heavy chains are each composed of variable and constant regions. The light chain has a V_L and C_L. There is a similar arrangement for the heavy chain (V_H and C_H). The Fab and Fc fragments may be isolated by proteolytic cleavage with the enzyme papain. The Fc fragments also contain two constant heavy chain regions (C_H2 and C_H3).

2. *IgM* is a receptor for antigen on B lymphocytes and is the first class of antibody made by a developing B lymphocyte.
3. *IgA* is the primary immunoglobulin found in saliva, milk, genitourinary, and tracheobronchial secretions and functions in external defense at mucosal surfaces. It is also known as secretory immunoglobulin (sIgA).
4. *IgD* is present on the surface of B lymphocytes exiting the bone marrow and targeted for 2° lymphoid organs. It is not a secreted immunoglobulin. It functions as an antigen receptor, but other functions for IgD are unknown.

5. *IgE* is associated with allergies and related diseases such as asthma. IgE mediates its effects through specific receptors on the surface of mast cells and basophils, causing the release of histamine (Chap. 13).

COMPLEMENT SYSTEM

The complement system is an important weapon in the first line of defense of the body against pathogens (innate immunity).

The complement system is an array of about 20 serum proteins activated by antibody binding to a pathogen (classic pathway) or directly by the pathogen (alternative pathway). The complement system facilitates inflammatory responses.

In either pathway, the pathogen becomes coated with complement initiating the complement cascade. The cascade includes the following:

1. Activation of the *membrane attack complex (MAC)* on the pathogen leading to perforations and lysis.
2. Production of *opsonins*, coatings to make antigens more palatable to phagocytes.
3. Release of *chemotactic agents (chemokines)* which attract phagocytes (*chemotaxis*) to the area of infection or inflammation.

ORGANS OF THE IMMUNE SYSTEM: BONE MARROW

The bone marrow is the source for all immune system cells. Cellular mediators of innate immunity travel directly from bone marrow through blood to sites of tissue damage. Immature T lymphocytes (pre-T cells) travel from bone marrow to thymus where they undergo education. B lymphocytes are educated in bone marrow, then travel to specific B-cell regions of lymphoid tissue. Macrophage precursors (monocytes) leave bone marrow and travel through blood to all organs and tissues where they undergo local functional specialization.

The blood-forming aspects of the bone marrow are discussed in greater depth in Chap. 17.

• B-LYMPHOCYTE EDUCATION

B-lymphocyte education results in a diploma consisting of the following immunologically relevant cell surface receptors: (1) IgM and IgD, (2) MHC class II proteins, (3) complement receptors, (4) immunoglobulin Fc receptors, and (5) co-stimulatory molecules. B-lymphocyte education also produces diversity in antigen recognition.

IgM and IgD are immunoglobulins found as membrane-bound antigen receptors on the surface of virgin B lymphocytes seeded to B-dependent areas of the lymph nodes, spleen, Peyer's patches, tonsils, and other lymphoid aggregations. It is important to realize that these immunoglobulin antigen-specific receptors develop in the bone marrow never having seen the antigen before. The MHC class II proteins are the molecules required for presentation of antigens and for eventual interaction with T lymphocytes to produce antibodies. The receptors for complement represent a link between antibody production and the complement system (serum proteins regulating the inflammatory cascade). The Fc receptor is specific for the Fc portion of antibodies. The finishing school for B lymphocytes occurs in the germinal centers of lymph nodes and other secondary lymphoid organs. This process of further education is discussed in the upcoming section on lymph nodes.

THYMUS

The thymus is the site of T lymphocyte education.

The thymus is located in the anterior mediastinum of the thorax and reaches its maximum size at about the time of puberty. It subsequently undergoes a process of involution in which *thymocytes* decline in number and fat infiltrates the organ. The combination of loss of cortical and medullary thymocytes coupled with infiltration of adipocytes blurs the boundary between the cortex and medulla in each lobule with increasing age. The histologic details of the thymus include a capsule and connective tissue divisions that form individual lobules with an outer cortex and inner medulla. T-lymphocyte education begins with the naive stem cells that arrive in the cortex from the bone marrow. As thymocytes migrate through the cortex toward the medulla, they proliferate and differentiate. Mature thymocytes that have avoided *programmed cell death (apoptosis)* and specific self-selection receive an exit visa if they have developed the appropriate surface markers.

• T-LYMPHOCYTE EDUCATION

The education process centers on the development of the full repertoire of T-lymphocyte antigen receptors and differentiation of T lymphocytes into either helper (**Th** or CD4+) cells or cytotoxic T (**Tc** or CD8+) cells. Instrumental in this process is the elimination of cells (clonal deletion) that recognize themselves as foreign.

The development of T-lymphocyte antigen receptors occurs through *gene rearrangement* as the cells proliferate and differentiate under the influence of the thymic microenvironment. This process is preprogrammed by the cell's genetic makeup and is completely antigen independent. T-lymphocyte antigen receptors contain binding sites for small segments (immunodominant peptides) of antigen molecules and binding sites for MHC class I or II molecules. The next step is the differentiation of T-lymphocyte subsets. T lymphocytes are defined as CD4+ by the presence of cell surface CD4 and CD8+ by the presence of cell surface CD8. CD4+ T lymphocytes recognize antigenic peptides bound to MHC class II molecules on the surface of antigen-presenting cells.

Some CD4+ T lymphocytes *(Th1)* provide assistance for CD8+ T-lymphocyte differentiation in response to antigen. Th1s themselves can also function as effector cells carrying out some forms of *cellular immunity*. Other CD4+ T lymphocytes *(Th2)* provide assistance for B-lymphocyte differentiation in response to antigen and are therefore involved in the generation of the *humoral (antibody-mediated) immunity*. CD8+ T lymphocytes recognize antigenic peptides bound to MHC class I molecules on the surface of antigen-presenting cells. The sole function of CD8+ T lymphocytes is as mediators of cellular immunity. Some CD8+ T lymphocytes (cytotoxic T lymphocytes, CTL) directly kill cells expressing foreign antigens. Other CD8+ T lymphocytes generate cellular immunity through secondary effector mechanisms.

During normal thymic education, T lymphocytes with the capacity for recognizing self-antigens are eliminated during development by a process known as *clonal deletion*.

The thymic T-lymphocyte repertoire includes cells that recognize antigens that they might encounter in their own body. If those cells were seeded into the periphery without any form of preselection, i.e., development of *self-tolerance*, they would function as "loose cannons." They would be able to mount an auto-immune response and attack self-cells. In fact, thymic selection sometimes is deficient, leading to development of diseases like type I (insulin-dependent) diabetes, rheumatoid arthritis, and Hashimoto's thyroiditis where pancreatic beta cells, the synovial villi, and the follicular cells of the thyroid, respectively, are attacked by T lymphocytes which see these self-tissues as foreign.

Educated CD4+ and CD8+ T lymphocytes exit from the thymic medulla and migrate to specific regions of the secondary lymphoid organs called T-dependent regions: *the deep cortex (paracortex)* of the lymph node, the *periarterial lymphatic sheath (PALS)* of the spleen, and the interfollicular regions of the Peyer's patches. These regions are discussed in detail in the section dealing with specific organ biology.

The education of T lymphocytes is shown in Fig. 19-2B. Pre-T lymphocytes leave the bone marrow for the thymus where they become naive thymocytes.

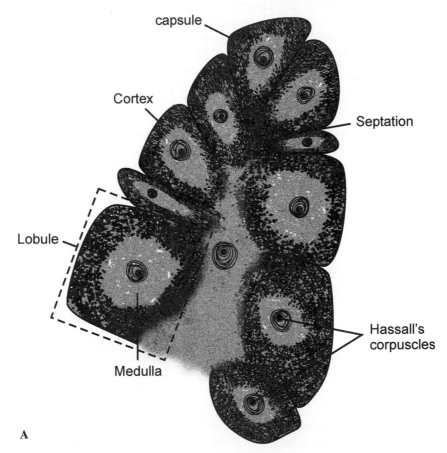

A

Figure 19-2 Histology of the thymus.
A. The histologic structure of the thymus showing the distinct lobulation with each lobule containing cortex and medulla.

Bone Marrow

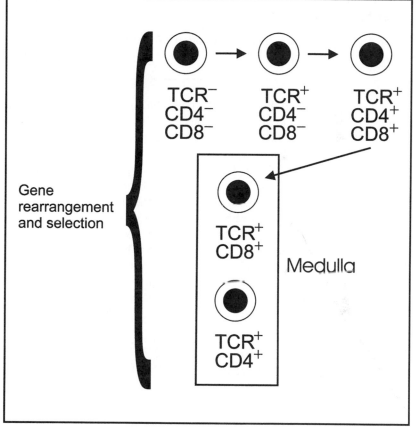

B

Figure 19-2 (cont.) T-cell education in the thymus.
B. Most of the education occurs in the thymic cortex in communication with epithelial cells and macrophages. Departure occurs from the medulla, although some cells may bypass the medulla completely. Errors in T-cell education, such as recognition of self as foreign, are destroyed in the thymus under normal conditions.

Over time, the T-cell receptor (TCR) matures: first as $TCR_{\gamma\delta}$ (immature form) then as $TCR_{\alpha\beta}$ (mature form). T cells in the periphery include less than 10 percent of $TCR_{\gamma\delta}$; the vast majority of circulating T cells are $TCR_{\alpha\beta}$-positive. T cells then develop as double positive cells (CD4+, CD8+) that differentiate into either the helper or cytotoxic phenotype.

LYMPH NODES

The lymph nodes function in active filtration of the lymph, but they also function as active filters of the blood.

    Lymph nodes are found throughout the body, but are found in highest concentrations in the axillary and inguinal regions as well as along the mesenteries. They are interspersed along the large lymphatic vessels and possess an elaborate system of sinuses for the filtration of the lymph.

    The structure of a typical lymph node is shown opposite. A capsule surrounds the entire lymph node. The lymph node has a distinctive bean-shape with a depression on one surface, called the *hilus*. This is the site of the efferent lymphatic vessels and the arteries and veins. The afferent lymphatics enter the convex side of the node. Lymph entering the lymph node proceeds from the afferent lymphatics through a tortuous series of sinuses formed by a meshwork of cells and fibers: subcapsular (marginal) to cortical to medullary. Macrophages, attached to the walls of these sinuses interact with cells moving through the sinuses, providing an excellent filtration system of phagocytic cells. (See Fig. 19-3.)

Lymphocytes leave the bloodstream through the high endothelial venules (HEVs) and target to specific regions of the lymph node.

    As shown in Fig. 19-4, lymphocytes circulate through the blood and those with specific receptors on the surface recognize *addressins* on the surface of endothelial cells of the high endothelial venule (HEV).

    Endothelial cells that are flattened, simple squamous cells normally line a venule. In the case of the HEV the lining cells are cuboidal. The binding of circulating cells to the HEV endothelium results in dissolution of the junctional complexes between the endothelial cells. This permits the lymphocytes to move between the endothelial cells (*diapedesis*) into the lymph node. The mechanism(s) involved in this process are discussed in the section on the vasculature

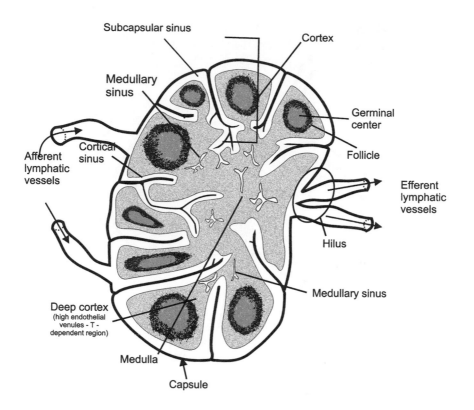

Figure 19-3 The histologic structure of the lymph node.
The lymph node filters the lymph and the blood. The cortex is the B-cell–dependent region and the deep cortex is the T-cell–dependent region. The sinuses (subcapsular, cortical, and medullary) are an essential part of the filtering system and extend from the subcapsular to cortical and eventually medullary sinuses.

and selectin-modulated infiltration of cells (Chap. 18). The high endothelial venules provide a unidirectional pathway for cells to enter the lymph node. B lymphocytes are directed to B lymphocyte–dependent (follicular) areas in the superficial cortex, whereas T lymphocytes are directed to the deep cortex (the region immediately around the high endothelial venules). Lymphocytes leaving the lymph node enter the venous system through the thoracic duct or the right lymphatic duct.

Lymph nodes produce antibodies and effector T lymphocytes.

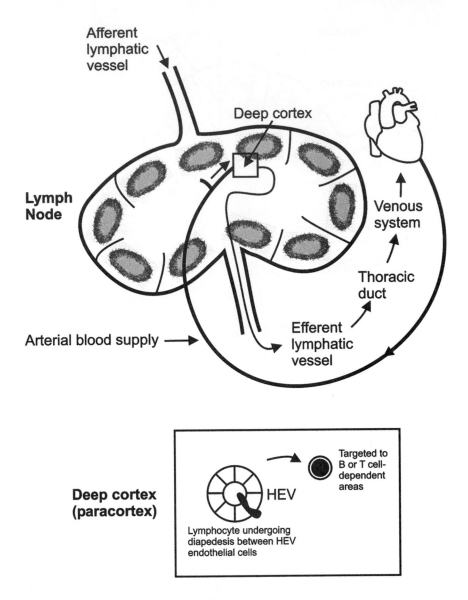

Figure 19-4 Lymphocyte recirculation.

The pattern of lymphocyte recirculation is shown in the figure. Passage of lymphocytes from the blood to the lymphoid tissue of the node is unidirectional and occurs through the high endothelial venules (HEVs) located in the deep cortex (paracortex) of the lymph node. Lymphocytes undergo diapedesis, a process by which they pass through the endothelial cell junctions of the HEV. Return of lymphocytes to the blood occurs via the thoracic duct and venous system and subsequently back to the circulation.

• **B LYMPHOCYTE RESPONSES TO ANTIGEN** The lymph node consists of an outer cortex and an inner medulla. The cortex contains germinal centers comprised of B lymphocytes, helper T lymphocytes, and macrophages. Exposure of an individual to antigen produces a B- and T-lymphocyte response that eventually leads to an acquired humoral immune response. Antigen presenting cells (APCs), derived primarily from the monocyte-macrophage lineage, phagocytose and process antigen wherever the individual happens to have been exposed. APCs include macrophages, dendritic cells, and even B cells. B cells function as APCs, taking up antigen (primarily of smaller, less complex antigens) by their surface immunoglobulin. The APC presents antigen to the Th2 lymphocytes that recognize antigen associated with MHC II on the APC surface and activates the TCR complex on the Th2 cell. This is one of two signals for Th2 activation. The second signal is provided by B7, a membrane-bound signaling molecule on the surface of the APC, which recognizes CD28 (a co-receptor protein on the Th2 cell). The activation of T cells is discussed in the next section. Once activated the Th2 lymphocyte expresses receptors for interleukin-2 (IL-2 receptors) and secretes IL-2, inducing cell proliferation of T cells that have encountered the specific antigen.

> Most B cells require Th2 lymphocytes for antibody production following antigen exposure. The interaction of Th2 (helper) and B lymphocytes requires two intercellular signals: 1) membrane-bound and 2) secreted.

All B cells require the presence of Th2 (helper) cells except for rare polymers with repeating antigenic determinants (*T-cell–independent antigens*). For T-cell–dependent antigens, intact antigen molecules stimulate B lymphocytes directly through antigen-specific cell surface immunoglobulin receptors. This is one signal required between the helper Th2 cell and the B cell. The other is expression and recognition of a transmembrane protein on the surface of activated T cells. When both signals are present, the Th2 cells release appropriate cytokines that stimulate B-lymphocyte growth and development. Antigen and activated Th2 lymphocytes (releasing cytokines) together stimulate activation, proliferation, differentiation, and class switching of antigen-specific B lymphocytes into plasma cells and memory B lymphocytes (Fig. 19-5). Plasma cells produce antibodies. The plasma cells move into the medulla where they secrete their antibodies into the bloodstream. Plasma cells do not normally enter the bloodstream. The production of high numbers of long-lived memory B lymphocytes that have the same antigen specificity as the original responding B lymphocytes means that there will be a much stronger antibody response following a second exposure to the same antigen. How humoral immune responses are generated in lymphoid organs is covered in considerably greater detail in *Basic Concepts in Immunology*.

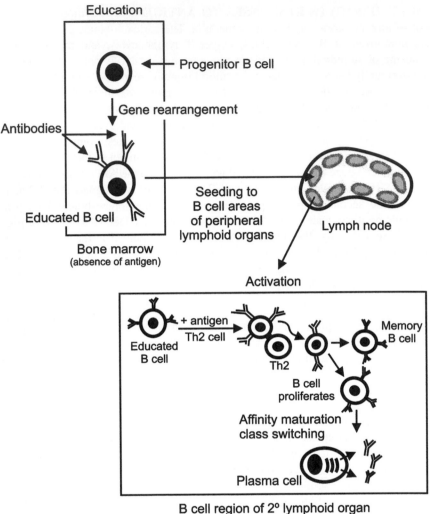

Figure 19-5 B-cell education and activation.
Education occurs in the bone marrow in the absence of antigen. In the presence of antigen, affinity maturation and class switching occur in B-cell regions such as the germinal centers of the lymph node, spleen, tonsils, Peyer's patches, and other lymphoid organs. Antigen activation converts a naive, educated B cell to an antibody-producing plasma cell and usually requires the presence of a helper T cell (Th2).

Fine adjustment of antibody production occurs during antigen exposure. The result is an increased antibody affinity (affinity maturation) that occurs through a mutational process called somatic hypermutation. The fine adjustment also includes class switching (isotype switching) in which B lymphocytes change the type of antibody (immunoglobulin) which they are synthesizing from primarily IgM to IgG, IgA, or IgE.

The affinity of antibodies increases with increased time after exposure to antigen. A process called *somatic hypermutation* (a dramatically increased number of point mutations) modifies the recombined antibody genes to produce higher affinity clones. High affinity clones have a competitive advantage in the presence of limiting concentrations of antigen. The result is selective expansion of the higher affinity clones. Memory responses then have higher relative affinity because higher numbers of higher affinity clones produce a higher amount of high affinity antibody.

B lymphocytes that have not been exposed to antigen in lymphoid organs possess IgD and IgM inserted on the surface as membrane-bound molecules. When stimulated by antigen in the presence of active T-lymphocyte help, some B lymphocytes undergo *class switching*. During class switching the immunoglobulin class expressed on the cell surface changes from IgM (the major antibody type of the primary response) to IgA, IgE, or IgG. Plasma cells then secrete a class of antibodies, IgM, IgA, IgE, or IgG, that is consistent with the switch. Memory cells also are products of class switching, possessing IgA, IgE, or IgG on their surfaces.

• T LYMPHOCYTE RESPONSES TO ANTIGEN

CD4+ and CD8+ lymphocytes and macrophages populate T-dependent regions of the lymphoid organs, such as the deep cortex of the lymph nodes.

Despite the presence of the T-cell receptor (TcR) and CD4 or CD8, seeded T cells are still naive. They must be primed before they can become effector T cells. Seeded T cells are antigen-ready, but require antigen stimulation (priming) before they can function in the immune response.

When an individual is exposed to antigen, macrophages that are specialized for antigen presentation phagocytose and process antigen wherever exposure occurs. The specialized antigen-presenting cells travel to a nearby lymphoid tissue where they present antigenic peptides to both CD4+ and CD8+ T lymphocytes. The resultant interaction activates both populations to express receptors for

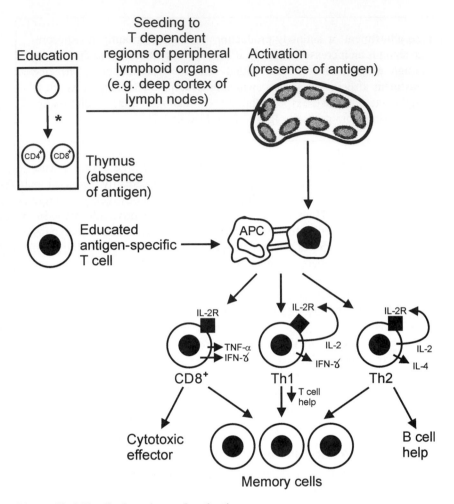

Figure 19-6 T-cell education and activation.
Progress from resting primed as a continuum influenced by antigen diversity. The asterisk delineates the education process (gene rearrangement and selection) shown in Fig. 20-2*B*. Expression of cytokines (IFN-γ, TNF-α, IL-2, and IL-4) is characteristic of priming. Effector cells are prepared for help and cytotoxicity.

interleukin-2 (IL-2 receptors) and the Th1 lymphocytes to produce interleukin-2 that stimulates proliferation of all T lymphocytes. This process is best thought of as *priming* of T cells (Fig. 19-6). The interaction of antigen-specific CD4+ Th1 lymphocytes with specialized antigen presenting cells produces proliferation and differentiation into primed CD4+ Th1 lymphocytes that provide help for priming CD8+ effector (cytotoxic) T lymphocytes. The primed T lymphocytes function as long-lived memory cells in secondary immune responses. In viral infections, primed CD4+ and CD8+ T lymphocytes acquire "exit visas," enter the circulation, and travel to sites of antigen exposure such as the upper respiratory tract. There they again are stimulated by antigen in association with specialized antigen-presenting cells. The primed CD4+ and CD8+ T lymphocytes differentiate into effector T lymphocytes that mediate cellular immunity. For example, during a viral infection, the effector lymphocytes may kill virus-infected cells, thereby limiting the ability of viruses to continue to spread and survive in the body. Killing is carried out through pore-forming enzymes called *per-forins*. An alternative pathway involves the secretion of *tumor necrosis factor-alpha* (TNF-α). Perforins act through a pathway of cytoplasmic cell death (*necrosis*), whereas TNF-α functions in programmed cell death (*apoptosis*).

Interferon-gamma (IFN-γ) is produced by cytotoxic T Cells (as well as Th 1 cells) and induces many of the genes associated with MHC including those for class I and class II (MHC). The overall effect is an augmentation of phagocytosis by macrophages and positive feedback on cytotoxic effector cells stimulating cell killing.

Cytokines such as IFN-γ are important because they activate effector T cells, increase the phagocytic capacity of macrophages, and induce expression of class II MHC on cells such as endothelial cells, which do not normally express this marker. The feedback loop between effector T cells and macrophages mediated by IFN-γ is essential in the destruction of antigens such as *Mycobacterium tuberculosis*, which can resist phagocytosis and live happily as an obligate parasite in nonactivated macrophages.

The essential difference between primed and effector cells is their expression of cytokines:

$$\text{primed cells} \begin{cases} \text{IL-2} \uparrow \\ \text{IFN-}\gamma \downarrow \end{cases} \qquad \text{effector cells} \begin{cases} \text{IL-2} \downarrow \\ \text{IFN-}\gamma \uparrow \end{cases}$$

In summary, T-cell differentiation involves a transition from primed cells to effector cells. Effector cells produce IFN-γ, whereas primed cells produce IL-2.

How cellular immune responses are generated in lymphoid organs and mediate their terminal effects is covered in considerably greater detail in *Basic Concepts in Immunology*.

MAJOR HISTOCOMPATIBILITY COMPLEX (MHC)

Major histocompatibility complex is a locus of genes that regulate tissue histocompatibility. MHC is a misnomer since transplantation between histoincompatible individuals is an artifical manipulation performed in clinical medicine.

The main function of MHC gene products is the presentation of antigenic peptides to T lymphocytes.

Class I and Class II MHC gene products are similar in structure (transmembrane heterodimers) and general function (binding peptides derived from foreign antigens during intracellular degradation, carrying the peptides to the cell surface and presenting them to T lymphocytes). Both class I and class II MHC gene products are members of the *immunoglobulin superfamily*. This is where the similarity ends. MHC class I is expressed on the surface of all cells except trophoblast and red blood cells, whereas MHC class II is expressed on the surface of B lymphocytes and antigen-presenting macrophages. There are actually distinct functions for class I and class II MHC molecules that are associated with specific cell types.

CD8+ T lymphocytes recognize peptide fragments of foreign proteins bound to MHC class I on the surface of cells.

CD8 is another member of the immunoglobulin superfamily. Both the CD8 protein and the T-lymphocyte antigen receptor are required for binding of the MHC class I protein fragment. Cells that are directly infected by parasites such as viruses express parasitic antigens on the cell surface. Antigen-presenting cells degrade antigens from the endogenous parasite into peptides that are released into the cytosol. Special transporters carry the degraded fragments from the cytosol into the lumen of the endoplasmic reticulum where they bind to specific receptors on MHC class I. The resultant peptide/MHC complex then is integrated into newly synthesized membrane and carried to the surface as part of normal membrane renewal. Once on the cell surface, the peptide/MHC class I complex antigen makes antigen presentation to T lymphocytes possible.

CD4+ T lymphocytes recognize peptide fragments of foreign proteins bound to MHC class II proteins on the surface of antigen-presenting cells.

Antigen-presenting cells also phagocytose or pinocytose antigens that travel to endosomes where they are enzymatically digested. Peptide digestion fragments combine with MHC class II protein. The peptide/MHC complex is delivered to the cell surface where it is recognized by the MHC class II dependent CD4+ T lymphocytes.

SPLEEN, TONSILS, AND MUCOSA-ASSOCIATED LYMPHOID AGGREGATIONS

The histology of these organs and the structure and function relationships of each organ are presented here. The germinal centers in each of these organs contain antigen-presenting cells, B lymphocytes, and helper T lymphocytes prepared to cooperate to produce antibodies on demand. The T-lymphocyte–dependent areas contain antigen-presenting cells, helper T lymphocytes, and effector T-lymphocyte (cytotoxic) precursors prepared to cooperate to produce cell-mediated immune responses.

SPLEEN

The spleen contains two histologically diverse areas: white pulp and red pulp dedicated to the functions of active blood filtration and recycling of red blood cell components, respectively.

The spleen is surrounded by a capsule and has blotchy areas of red and white in a fresh specimen. The diagram (Fig. 19-7) shows the structure of the spleen. In a H&E-stained section the appearance is blotches of red and purple representing the areas of red cell death and recycling (*red pulp*) and the areas of immune function with or without germinal centers (*white pulp*). The capsule of the spleen extends into the organ as trabeculae which carry blood vessels to and from outside the organ. The blood supply to the spleen passes from trabecular arteries that branch to form the central arteries (arterioles) of the white pulp. The arterioles continue to undergo reduction in size as they enter the red pulp. At this point there

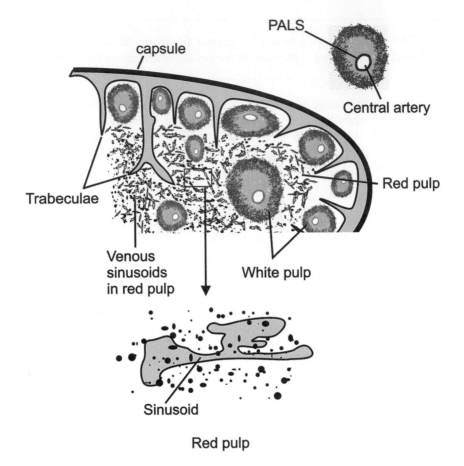

Figure 19-7 The histologic structure of the spleen.
Alternating areas (blotchiness) of red pulp (sinusoids with degenerating RBCs) and white pulp (areas of lymphocyte concentration) with prominent trabeculae extending from the splenic capsule are the histologic characteristics of the spleen.

has been debate over whether the splenic circulation is open or closed. In the open viewpoint, blood cells percolate from the small arterioles through the red pulp tissue before entering a venous sinus for return through the veins to the heart. The other viewpoint is that the splenic circulation is closed and blood flows from the small arterioles directly to the venous sinuses. The splenic circulation has some characteristics of each viewpoint. There is some tissue between the sinuses known as Billroth's cords, which are composed of capillaries, macrophages, and other wandering cells.

There are distinct T- and B-cell–dependent areas of the spleen.

In the spleen, B lymphocytes are located in the peripheral white pulp, whereas the T-dependent areas are found in the area surrounding the central artery near the center of the white pulp. That area is known as the *periarterial lymphoid sheath* (Fig. 19-7). The B-cell areas often have germinal centers and contain B cells undergoing their differentiation to plasma cells and memory B cells. T-cell responses are similar to those described previously.

Tonsils

The tonsils carry out immunologic surveillance of the head and neck.

There are three groups of tonsils: palatine tonsils, pharyngeal tonsils, and lingual tonsils forming Waldeyer's ring, a protective grouping of cells around the pharynx, near the union of the naso- and oropharynx. The type of epithelium that partially surrounds the tonsil can be used to determine its type (See Chap. 28). Crypts, deep invaginations of the surface, along with the partial encapsulation (or absence of encapsulation) are the hallmarks of the tonsils. The overall function of the tonsils is the monitoring, surveillance, and protection against viruses, bacteria, and foreign particles in the pharyngeal region. B- and T-cell regions function as they do elsewhere in the body.

• **PALATINE TONSILS**   Palatine tonsils are paired, partially encapsulated lymphoid organs anatomically positioned between the glossopalatine and pharyngopalatine arches. They are covered by a nonkeratinized, stratified squamous epithelium. Crypts extend from the surface and are also lined by the stratified squamous epithelium. A thin capsule of fibrous connective tissue covers the base and sides of the palatine tonsils.

• **PHARYNGEAL TONSILS**   Pharyngeal tonsils are located in the nasopharynx and are covered by a pseudostratified, ciliated epithelium typical of the respiratory system of which the nasopharynx is an essential part of the air conduction system. These tonsils are the "adenoids" that are frequently inflamed and hypertrophy, necessitating surgical removal in pediatric patients.

• **LINGUAL TONSILS**   The lingual tonsil is an accumulation of lymphoid tissue posterior to the circumvallate papillae within the substance of the tongue.

Mucosa-associated lymphoid tissue (MALT) and Peyer's patches

Mucosa-associated (MALT) has subsets of gut-associated (GALT) and bronchus-associated (BALT) lymphoid tissue. Peyer's patches are a key component of the GALT found in the ileum. M (microfold) cells are located in the epithelium overlying the dome of the Peyer's patch. The M cells sample antigens from the gut lumen. These antigens are phagocytosed by macrophages in the underlying lymphoid tissue and presented to neighboring T lymphocytes. The Peyer's patches produce antibodies in the B-dependent follicular areas. The interfollicular areas of the Peyer's patches are T-dependent–areas that contain helper T lymphocytes involved in maturation of IgA-secreting B lymphocytes as well as other types of B lymphocytes. Other T lymphocytes (CD8+) in the Peyer's patches are dedicated to a cytotoxic function.

FUNDAMENTAL PROCESSES—INNATE IMMUNITY

Innate immunity lacks specificity and memory.

In contrast to acquired immunity, innate immune protection against foreign agents is mediated by an inflammatory response that lacks antigen specificity and memory. The response is the same regardless of what the agent is that triggers it and regardless of the number of prior exposures to the agent.

Inflammation is the body's nonspecific response to infection or injury. It may be local or systemic and acute (short duration) or chronic (persisting for longer periods). It can be induced by injury alone or by foreign agents. The response is at the same time designed to repair the injury and to eliminate any foreign agents that enter the body as part of the injury. The reactive cells (and the foreign agents which stimulate them) are: neutrophils (bacteria), macrophages (viruses), and eosinophils (parasites).

The term *inflammation* is derived from "inflame," which means to produce heat. Heat production is only one of the characteristics of inflammation.

Inflammation consists of four classic characteristics: rubor (redness), tumor (swelling), calor (heat), and dolor (pain).

The redness is due to increased blood flow to the region. The swelling (edema) is due to increased capillary permeability. The heat and pain are due to local release of a variety of mediators.

Neutrophils and macrophages phagocytose invaders and release molecules that mediate further cellular recruitment and repair.

Neutrophils are the first immune cells to enter an inflammatory response, and their primary function is to phagocytose foreign invaders, particularly bacteria. They are soon followed by macrophages. Macrophages also phagocytose foreign invaders. Macrophages produce a cascade of local mediators that are responsbile for tissue repair. Chemokines (chemoattractant molecules), cytokines (small intercellular protein mediators), vasoactive amines, and growth factors mediate the cascade. There are complex symbiotic interactions between innate inflammatory responses and acquired immune responses that are beyond the scope of this book. Simply stated, however, inducing an inflammatory response is essential for efficient antigen presentation during generation of acquired immunity. In addition, the presence of an acquired immune response increases the intensity and efficiency of an inflammatory response against the agent that has induced the acquired immunity.

FUNDAMENTAL PROCESSES—ACQUIRED IMMUNITY

There are two types of acquired immune responses. One is the primary immune response that occurs following initial exposure to antigen. Secondary (memory) immune responses occur following each subsequent exposure to antigen. Secondary responses occur more rapidly, are maintained for longer periods of time, are quantitatively greater and are qualitatively more effective than primary responses. The differences between primary and secondary acquired immune responses can be explained by the three hallmarks of acquired immunity: diversity, specificity, and memory.

DIVERSITY

The diversity of acquired immune responses is explained by the fact that millions of clones of T and B lymphocytes, each of which expresses antigen receptors, imbue the immune system with distinct specificity.

Diversity is generated by antigen-independent rearrangement of antigen receptor genes during both T- and B-lymphocyte development and education.

Rearrangement occurs primarily in portions of antigen-receptor genes encoding antigen-binding sites. In T cells, the result of *gene rearrangement* is the production of T-cell receptors capable of recognizing a wide range of antigens. In the case of the B cells, the result is the production of antibodies specific for a large array of antigens. In the B-cell response, diversity is generated by rearrangement (*recombination*) of the gene segments that are related to structural components of the antibody: C (constant), V (variable), J (joining), and D (diversity). The process of rearrangment is similar in T cells, but involves the genes for the T-cell receptor instead of antibodies. The same recombination-activating enzymes are involved in the gene rearrangement required for the development of the T-cell receptor repertoire and antibody repertoire. Additionally, there is structural homology between T-cell receptors and antibodies that are both members of the immunoglobulin family.

The diversity generated by rearrangement occurs before exposure to antigen. In fact, it is essential that it not occur during antigen-induced clonal expansion when antigen specificity must be maintained.

This system of development and education has evolved so that lymphocytes exhibit a specificity repertoire that is capable of recognizing nearly all foreign pathogens and a wide range of other foreign agents. Diversity is generated in the absence of foreign antigen. In fact, if foreign antigens are introduced during T-lymphocyte development, the clones that recognize that antigen are eliminated in the same way that self-reactive clones are deleted from the repertoire. More detailed descriptions of the genetic events involved can be found in most standard immunology textbooks.

SPECIFICITY

Clonal selection is the process by which lymphocytes respond only to those antigens for which they have receptors. Clones (a family of cells that respond to similar antigenic determinants, i.e., epitopes) contain cells with identical antigen specificity.

Clonal selection explains the specificity of acquired immune responses in that antigens only stimulate those cells that have the appropriate antigen receptors on their surface. Antigens only stimulate proliferation and differentiation of those clones capable of recognizing that particular antigen. Mature lymphocytes maintain their identity when exposed to antigen. That is, all progeny of a particular clone have the same antigen specificity. One or more clones may be stimulated by an antigen, hence the terms *polyclonal* (more than one) and *monoclonal* that are often used to describe antibodies.

MEMORY

Memory is an essential part of the immune response and occurs in both cell-mediated (T lymphocytes) and humoral (B lymphocytes) immune responses.

Memory is explained partly by the fact that antigen exposure stimulates *clonal expansion*, thereby increasing the number of T or B lymphocytes capable of recognizing that particular antigen. *Memory cells*, the long-lived progeny of stimulated lymphocytes, are the foundation of immunologic memory. For example, in the case of an activated B lymphocyte, the cell proliferation results in an antibody-producing plasma cell and a memory B cell. Primed T lymphocytes function as long-lived memory cells in secondary immune responses. For example, in a tuberculin skin test, an extract of bacteria (tuberculin) is injected into the skin of an individual who has been exposed or previously immunized. The result is that the memory T cells that recognize and react to tuberculin secrete interleukins and induce an immune response at the site in the skin.

HOW THE ACQUIRED IMMUNE SYSTEM RESPONDS

Immune responses differ depending on a number of factors, but the key first step is the uptake of antigen near the entry site.

Immune responses vary qualitatively and quantitatively depending on the nature and amount of the antigen, the route of exposure, the condition of the host at the time of exposure and, of course, whether there has been a previous exposure. However, there are certain commonalities, and most antigens stimulate concurrent humoral and cellular immune responses. Regardless of which natural antigen induces the response, the first step is uptake by antigen-presenting cells at the site of entry. The immune response is enhanced if the agent or the type of damage at the same time produces nonspecific activation of the antigen-presenting cells. Macrophages or specialized cells derived from the monocyte and macrophage lineage (Langerhans cells in the skin, dendritic cells in the lymph nodes, microglia in the brain) are responsible for most antigen presentation. They are normal inhabitants of all tissues and are particularly numerous in tissues exposed to the external environment. Along with their inherent capacity for antigen presentation, macrophages have other important properties that are essential for generation of immune responses. They are mobile. After exposure to antigen they migrate from the exposure site to the closest lymphoid tissue where they interact physically with T lymphocytes exhibiting the appropriate antigen specificity. Soluble antigens released from the entry site into the blood also travel to lymphoid tissue where they bind to antigen receptors on B lymphocytes. Interaction of those three cell types leads to activation of B lymphocytes to divide forming antibody-producing plasma cells and memory B cells. The primary response leads mainly to production of IgM-type antibodies and IgM memory cells. If the antigen concentration is high enough, there will be some class switching and production, mainly of IgG and IgA antibodies and IgG and IgA memory cells. At the same time, antigen-presenting cells will combine with naive helper and effector T cells to stimulate a primary cellular immune response.

> The primary cellular immune response fundamentally involves increasing the number of specific memory T cells in lymphoid tissue throughout the body.

If the foreign agent is not highly pathogenic, as in common cold viruses, the primary immune response may ultimately eliminate the agent and the patient gets better. If the foreign agent is particularly pathogenic, as with the AIDS or smallpox viruses, the patient may have to be preimmunized to keep the agent from killing before being able to develop an effective immune response. For those who survive an initial exposure to the foreign pathogen, the fact that secondary responses develop much more rapidly and are more powerful will induce protection. Exposure results in a similar sequence of events that produces rapid elimination of the pathogen, often without producing any symptomatic evidence that exposure occurred at all. In medical practice, immunization stimulates primary responses and protects against pathogens that might be encountered in the environment. Although secondary immune responses represent the natural development of acquired immunity, immunizations make use of this memory system. Either a modified toxin or an attenuated virus can be combined with an immunologic adjuvant to produce a vaccine that remains antigenic (has the appropriate *epitopes* to stimulate the immune system), but is no longer pathogenic. The purpose of the adjuvant is to stimulate the activity of antigen-presenting cells at the vaccination site. The result is acquired immunity without the illness associated with the toxin or virus. In summary, lymphoid functions do the following:

1. Facilitate lymphocyte circulation and recirculation through the lymph and blood to carry out surveillance and attack on antigens
2. Transport activated effector lymphocytes and antibodies to their site(s) of action
3. Concentrate antigens in secondary lymphoid organs where immune responses occur
4. Eliminate antigen

RESPIRATORY SYSTEM

·

- **Typical Respiratory Epithelium**
- **Nose and Nasal Cavities**
- **Nasopharynx**
- **Larynx**
- **Trachea and Bronchi**
- **Bronchiole**
- **Alveoli and Alveolar Septa**

·   ·   ·   ·   ·   ·   ·   ·   ·   ·   ·   ·

The respiratory system consists of a conducting portion and a gas exchange portion. The function of the former is to deliver warm, moist, clean air to the gas-exchange portion. The function of the gas-exchange portion is to provide the most efficient mechanism for getting O_2 into the blood and CO_2 out. The histologic structure of both regions mirrors their functions. In addition, as in other organs, the associated nervous and connective tissues, as well as immune defenses, play an important role in maintaining the health and efficiency of the respiratory system.

TYPICAL RESPIRATORY EPITHELIUM

The generalized respiratory epithelium is a *pseudostratified, ciliated columnar epithelium.* In addition to ciliated columnar cells, the typical respiratory epithelium contains stem cells, goblet cells, and neuroendocrine cells. Neuroendocrine cells are found individually as in other organ systems, but also in groupings called neuroepithelial bodies which may function as sensory receptors and are more prevalent in infants. Connective tissue cells and serous and/or mucous glands may be found in the submucosa (Fig. 20-1). A thin zone of loose connective tissue (*lamina propria*) is found immediately beneath the epithelium. It blends imperceptively into the subjacent *submucosa* consisting of dense irregular connective tissue. Basal cells of the pseudostratified epithelium can only replace themselves. The mucous (goblet) cell appears to be the stem cell which can differentiate into the other cells that make up the epithelium.

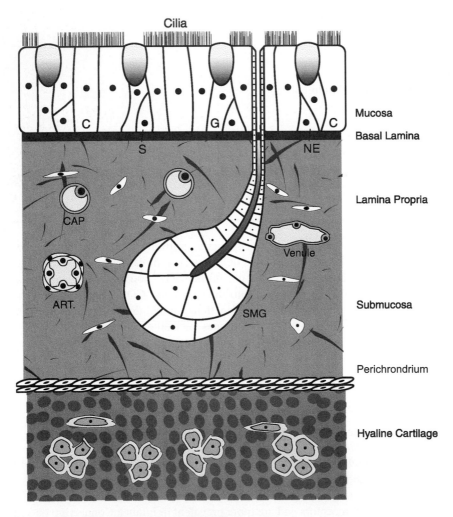

Figure 20-1 General respiratory epithelium.
The pseudostratified, ciliated columnar epithelium contains ciliated columnar cells (C), goblet cells (G), stem cells (S), and neuroendocrine cells (NE). Ducts from seromucous glands (SMG) reach the lumenal surface.

The exact nature of the respiratory epithelium varies from region to region depending on functional requirements.

NOSE AND NASAL CAVITIES

The nares (openings of the nostrils) are marked by a stratified squamous epithelium continuous with the epidermis of the face and slowly converting to a typical respiratory epithelium (pseudostratified, ciliated columnar) at the level of the nasal septum.

In addition to sensory functions, the primary function of the nasal cavity is to warm, clean, and humidify inspired air.

Underlying the epithelium in most areas of the nasal cavity is a rich capillary bed close to the surface which transfers heat from the blood to the inspired air. Numerous mucous glands add moisture. The nasal conchae (turbinate bones) increase the surface area of the nasal epithelium, create air turbulence, and increase the warming and humidifying capacities of the nasal cavities. A venous plexus (*swell body*) on each side beneath the epithelium of the inferior and middle conchae regulates airflow through the nasal cavities and allows the epithelium in each nostril to rest on an alternating hourly schedule.

The *paranasal sinuses* are lined by a simplified respiratory epithelium which, for the most part, is simple, columnar, and ciliated.

A feature of the stratified squamous epithelium of the nares is the presence of thick hairs, or vibrissae.

The *vibrissae* are the first part of the defense mechanism of the respiratory system and filter out dust and other large particles from the air. They have a propensity to increase in length and number with age along with the hair in the ears.

Two regions of the nasal cavity contain a specialized sensory epithelium—the olfactory mucosa and the vomeronasal organ.

The *olfactory epithelium* (Fig. 20-2) occupies the roof of the nasal cavity adjacent to the superior concha/meatus on each side. In addition to the cells of the general respiratory epithelium, the olfactory epithelium contains *specialized receptor cells for odors.*

The olfactory receptor cells are true bipolar neurons with an enlarged apical region (olfactory vesicle) from which emanate large, immotile cilia studded with G-protein–linked (G_{olf}) odor-specific receptors.

For a review of G-protein–linked receptors, see Chap. 9. The axonal processes of the olfactory neurons form the *fila olfactoria*, which pierce the cribriform plate of the ethmoid bone to synapse in the olfactory bulbs. *The receptor cells senesce and are replaced from the stem cells*, a phenomenon unique to olfactory receptor cells and some populations of hippocampal neurons. Acini of *serous glands* (Bowman's glands) are found in the underlying connective tissue, and their ducts reach the lumen (airway). Their proteinaceous products dissolve chemical odorants, which facilitates their presentation to the olfactory receptors. (See Chap. 27 for more information on signal transduction in olfaction.)

The vomeronasal (Jacobson's) organs (VNO) are located in the respiratory epithelium on either side of the nasal septum and are histologically similar to the olfactory epithelium. They are thought to respond to sexually oriented stimuli in the form of pheromones.

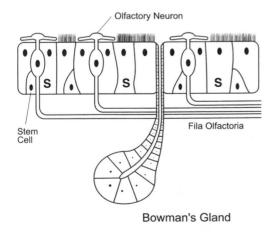

Figure 20-2   Olfactory epithelium.
The specialized olfactory neurons possess modified cilia with receptors for odorants. Their axons make up the fila olfactoria of the olfactory nerve. Ciliated columnar supporting cells (S) and stem cells make up the remainder of the epithelium.

NASOPHARYNX

The nasopharynx links the nasal cavities with the oropharynx. The respiratory epithelium of the nasal cavity continues into the oropharynx and gradually transforms into the stratified squamous, nonkeratinized epithelium which lines the oropharynx.

Mucosa-associated lymphoid tissue (MALT) is abundant in the nasopharynx. It is the first line of immunologic defense in the respiratory system.

The disposition of lymphoid tissue around the nasopharynx is termed *Waldeyer's ring*. The ring includes the *pharyngeal tonsils (adenoids)* at the juncture of the roof and posterior wall of the nasopharynx. In the tonsils and other MALT, macrophages attack foreign substances, portions of which are presented to lymphocytes to initiate humoral and cellular immune responses. (See Chap. 19.)

LARYNX

The larynx is located between the pharynx and trachea. It is supported by cartilages linked together by ligaments and muscles. The larynx (specifically the glottis) functions as a valve to prevent swallowed food from entering the respiratory tract. In humans, the larynx also functions as a critical participant in phonation and the production of sound. Whether the singer is Mick Jagger, Celine Dion, Placido Domingo, or an amateur in the shower, voiced sound is produced by vibration of the vocal cords with resonance produced by the laryngeal cavities. In the study of histology, a good slide of the larynx is worth its weight in gold. It contains two types of epithelium, skeletal muscle, two types of cartilage, and a variety of glands, as well as numerous blood vessels, peripheral nerves, (loose, dense irregular and dense regular) connective tissues, and even some bone.

Both typical respiratory epithelium and stratified squamous epithelium are found in the larynx.

The lingual and upper laryngeal portions of the epiglottis are covered by nonkeratinized stratified squamous epithelium as are the *true vocal folds*, bilateral median ridges containing the vocalis muscles. Stratified squamous epithelium generally occurs where the surface is susceptible to abrasive forces. The true vocal folds are covered by this type of epithelium due to the motion and velocity of the air passing over them. The vocal ligament and the vocalis (skeletal) muscle are found deep in the fold. The remainder of the larynx is lined by pseudostratified, ciliated columnar epithelum.

Both hyaline and elastic cartilages are found in the larynx.

The elastic cartilages include the epiglottis, the cuneiforms, the corniculates, and the tips of the arytenoids. The thyroid, cricoid, and remainder of the arytenoids are hyaline cartilage. Remember from Chap. 14 that the elastic fibers coarsing in the matrix between chondrocytes in elastic cartilage are best visualized with *silver or orcein stains*.

The bone found in some sections of the larynx is the hyoid bone.

In H&E sections, the bone stains bright pink. In optimal laryngeal sections, the hyoid bone presents as a circular or oblong profile with prominent bone marrow. Megakaryocytes and blood cell precursors are often readily identifiable here. (See Chap. 17.)

Other histologic features of the larynx include the following: skeletal muscle (intrinsic and extrinsic muscles of the larynx) with tendinous (dense regular connective tissue) attachments to the cartilage; seromucous glands in the subepithelial dense irregular connective tissue; isolated lymphoid nodules; blood and lymphatic vessels; and medium-small size nerve fascicles which are motor and sensory branches of the vagus nerve (cranial nerve X).

TRACHEA AND BRONCHI

The trachea, bronchi, and bronchioles are passageways for air moving toward and away from the portions of the lungs where gas exchange occurs. They also provide a mechanism for the removal of pollutants (the tracheobronchial escalator).

The lumen of the trachea is lined by a typical respiratory epithelium. In humans, a distinct lamina propria, submucosa, and outer adventitial covering can be observed. Elastic fibers are prominent in the lamina propria.

> A distinctive feature of the trachea is the presence of C-shaped cartilage rings.

A series of 16 to 20 C-shaped rings of *hyaline cartilage* separate the submucosa from the adventitia. They are covered by a perichondrium which blends into the surrounding connective tissue. (See Chap. 14.) The cartilage rings are deficient posteriorly where the gap is filled by a fibroelastic band and *smooth muscle cells (trachealis muscle)*. The cartilages prevent the collapse of the trachea during forced inspiration, whereas the muscle and fibroelastic band moderate the diameter of the trachea slightly.

> The respiratory system, like the gastrointestinal tract, is particularly susceptible to invasion of microbes from the external environment. It therefore has its own system of immune defenses.

The system within the respiratory tract is known as the BALT (bronchus-associated lymphoid tissue) and contains T and B lymphocytes and other immune accessory cells discussed in the lymphoid chapter. (See Chap. 19.)

> The cilia of the bronchioles, bronchi, and trachea move mucus and entrapped particles, cellular debris, and dissolved substances toward the oropharynx.

The cilia and mucus secreted by goblet cells and glands of the bronchioles, bronchi, and trachea form the *tracheobronchial (mucociliary) escalator* which moves dust and other particles out of the lungs and toward the oropharynx where they may be expectorated or swallowed. Similarly, the cilia of the respiratory epithelium of the upper respiratory tract (nasal cavities and nasopharynx) also beat toward the oropharynx.

> Immotile cilia result in inability to clear mucus from the respiratory passages.

As noted earlier, cilia are an important component of the mucociliary escalator which moves mucus and entrapped contaminants toward the oropharynx. The structure of cilia and the mechanisms of ciliary movement are discussed in Chap. 8. Cilia move by bending an intricate bundle of microtubules and associated proteins (axoneme). Movement of the microtubules that make up the ciliary axoneme is mediated by the molecular motor, ciliary dynein. Genetic defects in the gene coding for ciliary dynein can result in *immotile cilia (Kartagener's) syndrome.* Of course, such genetic defects would result in immotile cilia in other systems with ciliated cells such as the male (sperm) and female (cells of the oviduct) reproductive systems. Although such problems could result in infertility, the interference with the function of the respiratory system is more life-threatening.

The bronchi are characterized by a typical respiratory epithelium and the presence of cartilage plates which diminish in size distally.

The primary or main bronchi branch from the trachea and, for a short distance, maintain its histologic organization. Eventually (1) the cartilage rings become disrupted into cartilaginous plates, which diminish in size and number distally, and (2) smooth muscle invades the space between the submucosa and the cartilage to form a spirally oriented latticework around the smaller bronchi and bronchioles.

BRONCHIOLE

The bronchioles are the last part of the conducting portion of the respiratory system. They include conducting bronchioles, terminal bronchioles, and respiratory bronchioles.

The bronchiolar epithelium progresses from a simple columnar ciliated epithelium proximally to a cuboidal epithelium distally. Airflow through the bronchioles is controlled by smooth muscle.

In addition to the gradual change in the epithelium, bronchioles are characterized by the *absence of cartilage, absence of glands,* and *only sparse goblet cells.* Smooth muscle occupies a large relative portion of the bronchiolar wall. The smooth muscle of the respiratory tract is innervated by the autonomic nervous system and is also regulated locally by secretions of neuroendocrine cells. *Sympathetic innervation relaxes bronchiolar smooth muscle and thus dilates the airways*. Many asthma medications are sympathomimetic drugs. Parasympathetic input results in constriction of the airways.

Clara cells are found only in the bronchioles.

Clara cells are a specialized epithelial cell characteristic of the bronchioles. The apical portion of the Clara cell is dome-shaped and bulges into the airway lumen. Clara cells *secrete protein components of surfactant* (see below) and contain abundant smooth endoplasmic reticulum (ER) with oxidative enzymes such as cytochrome P_{450} and thus may *play a role in local detoxification*. P_{450} is discussed in more detail in relation to the liver (Chap. 22).

Proper composition, consistency, and function of airway mucus depends on the activity of epithelial ion channels.

Cystic fibrosis (CF) is a disease resulting from genetic defects in the gene coding for a transmembrane chloride channel. (Review transmembrane proteins in Chap. 1.) In many of these cases, the defect causes the deletion or substitution of a single amino acid which results in *incorrect folding* of the newly synthesized channel protein (*cystic fibrosis transmembrane conductance regulator, CFTR*). Incorrectly folded proteins are destroyed in the rough ER (Chap. 5). Therefore, these channel proteins never reach the cell membrane, and the cells are unable to transport chloride ions efficiently. In the respiratory system, this results in the inability of cells of the epithelium to transport chloride out of the cells into the lumen of the airway. Sodium remains with the chloride, and water is drawn out of the mucus. This results in a sticky mucus which attracts bacteria and clogs airways because it is not easily moved by the tracheobronchial (mucociliary) escalator. DNA resulting from the breakdown of cells adds to the stickiness of the mucus. For this reason, inhalation of DNAase is currently used in the treatment of cystic fibrosis. Effects of CF in other tissues are discussed in Chap. 22.

Terminal bronchioles are the last and smallest division of the respiratory tract involved solely in air conduction. Some gas exchange occurs in the respiratory bronchioles.

Terminal bronchioles are lined by a cuboidal ciliated epithelium. They branch into respiratory bronchioles, which are the initial segments of the gas-exchange portion of the respiratory system. *Respiratory bronchioles* are characterized by *occasional alveoli* in their walls. Isolated spiral bundles of smooth muscle underlie the epithelium. Respiratory bronchioles branch or transition into alveolar ducts which consist of numerous outpocketings of alveoli interspersed with small bundles of smooth muscle. The ducts end in alveolar sacs which are formed from the openings of several small groups of alveoli (Fig. 20-3).

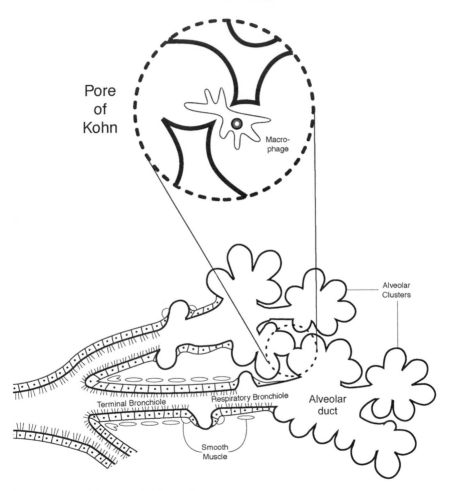

Figure 20-3 Distal bronchioles and alveoli.
Terminal bronchioles with a ciliated simple cuboidal epithelium lead into one or more respiratory bronchioles characterized by isolated alveoli. Respiratory bronchioles transition into alveolar ducts which open into alveolar clusters. Openings (pores of Kohn) between adjacent alveolar clusters allow equalization of air pressure and alternative pathways for airflow and alveolar macrophages.

ALVEOLI AND ALVEOLAR SEPTA

The walls of the alveoli and alveolar septa are lined by a thin epithelium consisting primarily of *type I alveolar cells* (type I pneumocytes). Although these cells are not the most numerous in the alveolar wall, they cover most of the surface area. *Type II alveolar cells* (type II pneumocytes) are cuboidal and serve as a stem cell for both type I and type II cells (Fig. 20-4). Alveolar macrophages (dust cells) arise from monocytes like other phagocytic cells of the body.

Surfactant increases the efficiency of inspiration and serves a protective function.

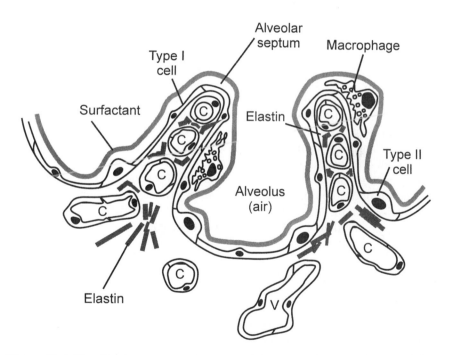

Figure 20-4 The alveolar septum.
The alveolar lining consists of flattened type I pneumocytes which function in gas exchange and cuboidal type II pneumocytes, which secrete surfactant. Alveolar macrophages patrol the alveolar surface beneath the lipid layer of surfactant. The blood-air barrier consists of the thin alveolar epithelium, the endothelium of capillaries in the septal wall, and the intervening shared basal lamina (C, capillary; V, venule).

Type II pneumocytes contain numerous *lamellar bodies* containing phospholipids, particularly *dipalmitoyl phosphatidylcholine (lecithin)*, the primary lipid component of *surfactant*. The large phospholipid whorls are distinctive and make the type II pneumocyte one of the few cell types which students should be able to specifically identify at the electron-microscopic (EM) level. When secreted from the apical aspects of these cells, the lipid, along with an underlying aqueous phase containing proteins secreted by other cells (e.g., Clara cells), forms a thin layer on the alveolar surface which *reduces surface tension* and allows effective inspiration. Surfactant molecules are interspersed with water molecules, reducing stickiness and surface tension at the air-fluid interface. Surfactant also balances the intraalveolar pressures between large and small alveoli. The aqueous phase of surfactant contains proteins with antibacterial and antiviral properties. Old surfactant is phagocytosed by the type II pneumocytes and alveolar macrophages.

During development, surfactant is essential for the adaptation to the extra-uterine environment.

The respiratory system must be ready to function at birth. Premature infants younger than 7 months gestation are particularly prone to *respiratory distress syndrome* (formerly called hyaline membrane disease) because their type II pneumocytes are immature and unable to produce sufficient quantities of surfactant, resulting in poor inspiration. Often, the level of surfactant production can be assessed *in utero* by amniocentesis. Current treatments include intratracheal administration of natural or synthetic surfactant to the newborn (replacement therapy) or administration of glucocorticoids (e.g., betamethasone) to the mother at least 24 h prior to birth. The glucocorticoids traverse the placenta and stimulate fibroblasts in the lungs of the fetus to secrete a factor which stimulates the synthesis and secretion of surfactant by type II pneumocytes.

Macrophages are found on the alveolar surface beneath the surfactant layer.

Alveolar macrophages, also called dust cells, patrol the alveolar surface and ingest any particles which escaped the tracheobronchial escalator, including pollutants, bacteria and, as noted above, surfactant. In cases of congestive heart failure (CHF), fluid containing breakdown products of hemoglobin (iron-containing hemosiderin) leaks into the alveolar air space and is phagocytosed by the alveolar macrophages. These iron-containing macrophages are termed *heart failure cells* and are diagnostic for CHF. Alveolar macrophages also play a role in the etiology of *emphysema*. (See below.)

Gas exchange occurs across the blood-air barrier.

In the septa between alveoli, the capillary walls (Fig. 20-4) come into close contact with the thin epithelium of type I pneumocytes. The *blood-air barrier* consists of the thin capillary endothelium (a continuous endothelium), the thin pneumocyte epithelium, and an intervening basal lamina contributed by both cell types. This minimal barrier permits the exchange of gases but does not allow the passage of fluid or cells into the alveoli under normal conditions. O_2 diffuses across the barrier from the air to the RBCs in alveolar capillaries where it is bound to hemoglobin. CO_2, formed by the breakdown of H_2CO_3 by carbonic anhydrase in RBCs, diffuses in the opposite direction.

The connective tissue of the alveolar septum is especially rich in elastin, which aids in expiration.

Elastin is an important component of the alveolar septum, since it permits the *elastic recoil* of the alveoli after they are inflated during inspiration, thus aiding in expiration. *Emphysema* (difficulty in expiration) occurs when macrophages and neutrophils are chronically stimulated to release the enzyme *elastase*, which degrades *elastin* and thus decreases the recoil capacity of the alveolar wall, eventually leading to breakdown of the interalveolar septum. Although cigarette smoking and air pollution are the primary causes of emphysema, a genetic defect or deficiency in α_1-*antitrypsin*, an enzyme which degrades other enzymes, also results in an increase in elastase leading to emphysema. Emphysema, along with asthma and chronic bronchitis, is classified as chronic obstructive pulmonary (airway) disease (COPD), the most common lung disease in humans.

Cigarette smoking and environmental air pollution are carcinogenic.

Emphysema is one disease which results from cigarette smoking and environmental pollutants. In addition, squamous cell carcinoma, the most common lung tumor, induces changes in the bronchiolar epithelium leading to the transformation of the pseudostratified epithelium to a stratified squamous type (hence the term *squamous cell carcinoma*).

The lungs receive a dual blood supply from both the systemic and pulmonary circulations.

The alveoli of the lungs receive blood (O_2-poor) only from the pulmonary circulation. These vessels tend to follow closely the distribution of the airways. The parenchyma and visceral pleura of the lungs receive nutrition from systemic arteries. These are generally recognizable by their more muscular tunica media compared with similarly sized pulmonary vessels.

GASTROINTESTINAL TRACT

•

- **Oral Cavity**
- **Role of Saliva in Digestion**
- **Teeth**
- **Tongue**
- **Layers of the Digestive Tube**
- **Esophagus**
- **Stomach**
- **Small Intestine**
- **Digestion**
- **Large Intestine**
- **Cell Turnover**
- **Enteroendocrine Cells**
- **Gut-Associated Lymphoid Tissue (GALT)**

•　　•　　•　　•　　•　　•　　•　　•　　•　　•　　•　　•

The gastrointestinal (GI) tract is a hollow tube that passes through the body from oral cavity to anus, but is never in the body. The gastrointestinal tube has two types of linings:

1. Stratified squamous epithelium for protection at the two ends and through some of the frictional passageways (oral cavity, esophagus, and anus)
2. Simple columnar epithelium modified into glands for different functions in the stomach, small intestine, and large intestine

The external muscle also varies along the length of the tube:

1. Skeletal muscle at the two ends (oral cavity, tongue, pharynx, and upper parts of the esophagus, returning at the anus)
2. Smooth muscle throughout most of the GI tract (lower part of the esophagus, stomach, small intestine, and large intestine)

The GI tract possesses an intrinsic rhythmicity, which is largely due to the existence of an enteric nervous system.

Although there are inputs from the parasympathetic (vagus nerve) and the sympathetic nervous systems, the intrinsic nervous system exists independently of these external influences. A variety of peptide neurotransmitters function in this system: bombesin, motilin, vasoactive intestinal peptide, and many others. Unicellular enteroendocrine cells of the epithelium also synthesize these peptides. The endoderm is the source of the enteroendocrine cells, unlike the intrinsic nerve fibers, which are derived from neural crest. Vagal nerve fibers stimulate the movement of material through the GI tract by a process called peristalsis, which involves activation of the external muscle layers to move food through the system.

The vulnerability of the GI tract to invading microorganisms results in an elaborate system of mucosa-associated lymphoid tissue (MALT).

In the case of the GI tract, it is called GALT (gut-associated lymphoid tissue). This system has greater surface area to protect than the SALT (skin-associated-lymphoid tissue), which protects the outer surface.

The GI tract is a victim of abuse. Humans eat an array of foods, some with powerful spices. Many humans drink ethanol to excess and gorge themselves with "all-you-can-eat buffets," or greasy high-fat meals. The poor GI tract only rarely receives appropriate care and feeding.

The GI tract protects itself in two major ways: (1) the production of large amounts of mucus to assist in lubrication and reduction in friction and (2) the rapid turnover of epithelial lining cells in the most harsh environments like the stomach and small intestine.

ORAL CAVITY

The mouth (oral cavity) is established early during embryonic development, through the breakdown of the oropharyngeal membrane, allowing the embryo to swallow the amniotic fluid. The mouth experiences a tasty diet of amniotic fluid and urine in fetal life and all the delights of tasting (corned beef, burgers, enchiladas, and Hunan ginger beef) in childhood and adulthood. The oral cavity contains the teeth and the tongue (including the taste buds).

ROLE OF SALIVA IN DIGESTION

Digestion begins in the mouth with the action of saliva containing mucus and amylase. It is generally believed that the food is in the mouth an insufficient time for significant digestion to occur. The three major salivary glands produce a large quantity of saliva. The *parotid gland* produces a proteinaceous (*serous*) secretion, the *sublingual gland* secretes *mucus*, and the *submandibular gland* has a *mixed secretion*, both serous and mucous. Minor salivary glands are found throughout the oral cavity and include glands of all three types: serous, mucus, and mixed. (See Chap. 22.)

Saliva has a number of other functions including (1) moistening the oral cavity; (2) cleansing the mouth; (3) moistening the food and facilitating the forming of a bolus: (4) facilitating swallowing; (5) permitting the appreciation of taste; (6) antibacterial influence through lysozyme, RNAse, DNAse, and secretory immunoglobulin (IgA); and (7) a role in fluid balance.

TEETH

The calcified tissues of the teeth include enamel, dentin, and cementum. The soft tissues of the tooth include the pulp, periodontal ligament, and gingiva (see Fig. 21-1).

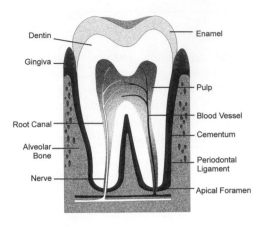

Dentin

Gingiva

Root Canal

Alveolar Bone

Nerve

Enamel

Pulp

Blood Vessel

Cementum

Periodontal Ligament

Apical Foramen

Figure 21-1 The structure of an adult molar tooth.

Three embryonic tissues (ectoderm, neural crest, and mesoderm) interact in the development of the teeth.

The enamel organ (those cells committed to the formation of the enamel) is formed from the ectoderm under the inductive influence of the mesoderm and neural crest. This is one of the epithelial-mesenchymal interactions that are very important in development of many body organs. Neural crest and mesoderm give rise to the dental papilla that forms the dentin, cementum, and pulp as well as surrounding connective tissue and periodontal ligament (Fig. 21-1).

The baby, or deciduous, teeth erupt after birth. The dentition follows the formula 2:1:2 (molars:canines:incisors) for each quadrant, with a total of 20 deciduous teeth. These teeth are replaced during childhood. The permanent teeth have the formula 3:2:1:2 (molars:premolars:canines:incisors) in each quadrant, with a total of 32 teeth.

The dentin is similar to bone except harder, owing to increased Ca^{2+} content. Odontoblasts synthesize dentin.

The *odontoblasts* line the dental pulp as a single layer of cells that produce *predentin* (analogous to prebone formed by osteoblasts). Predentin subsequently mineralizes to form *dentin* through the deposition of hydroxyapatite on a collagenous matrix.

> Enamel covers the crown of the tooth, is the hardest substance in the body (highest Ca^{2+} content), and is synthesized by ameloblasts.

Unlike other calcified tissues, enamel contains an organic matrix composed of amelogenins and enamelins rather than collagen. However, the organic components, along with water, are removed after calcification to leave an enamel that is >97 percent inorganic. The last of the calcified tooth tissues, *cementum*, is similar to bone and covers the dentin of the roots of the tooth. The teeth reside in sockets within the alveolar bone of the maxilla and mandible (upper and lower jaws, respectively).

The noncalcified tissues of the tooth are important parts of the tooth. They contain the nerves and blood vessels and provide a means for limited movement and adaptation.

> The tooth pulp is derived from mesoderm and is both highly vascular and innervated.

The primary cell type of the pulp is the fibroblast. Odontoblasts differentiate from cells in the pulp.

> The periodontal ligament is a highly metabolically active tissue, binds the cementum of the tooth to the bony socket, and allows limited movement.

The periodontal ligament absorbs pressures exerted by mastication and prevents transduction of pressure directly to the alveolar bone. Orthodontic movements are possible because of the ability of the periodontal ligament to adapt to change. The periodontal ligament is affected in diseases such as diabetes mellitus. Scurvy also affects the periodontal ligament because of its high rate of collagen synthesis. Scurvy is caused by vitamin C deficiency. Humans, as well as guinea pigs, require vitamin C because they cannot synthesize it from glucose as do other species. Deficiency in vitamin C leads to impairment of peptide hydroxylation in collagen. Nonhydroxylated collagen cannot form the proper triple helix leading to loss of teeth. Remember the classic picture of the toothless British sailor from the 1880s. The term *limey* for British sailors is derived from the use of lime juice to provide vitamin C (ascorbic acid) and prevent scurvy.

The *gingiva*, or *gums*, are the soft tissues associated with the tooth. The gingiva consists of a stratified squamous epithelium and the underlying connective tissue, which surrounds the tooth like a collar. Its contact to the tooth is called the epithelial attachment and is critical for the health of the gingiva. The gingiva is affected in diseases such as diabetes mellitus.

TONGUE

The tongue is a mass of skeletal muscle covered by a stratified squamous epithelium containing elevated specializations called *papillae*. There are four types of papillae: *filiform* (conical-shaped), *fungiform* (mushroom-shaped), *foliate* (leaflike folds), and *circumvallate* (large, circular, surrounded by a moat). Filiform papillae are the only type which contains no taste buds, and the foliate papillae are rudimentary in adult humans. The taste buds contain supportive (sustentacular) and taste cells, both derived from a single stem cell. The taste cells possess apical microvilli that contain the taste receptors. Afferent fibers are located close to the base of the taste cell; they are activated by neurotransmitter release from the basal part of the cell following an increase in intracellular Ca^{2+}.

The taste buds can detect four basic taste sensations. Different sensations have a regional distribution along the tongue surface: salty and sweet at the tip of the tongue, acid (sour) at the sides, and bitter in the area of the circumvallate papillae. Chemicals producing taste sensations are dissolved in saliva or the serous secretions of von Ebner's glands whose ducts empty into the moats at the bases of the circumvallate papillae.

The four taste sensations utilize a variety of signal transduction pathways leading to depolarization of the taste cell in all cases except bitter taste. Bitter taste involves a hyperpolarized state similar to retinal cell signal transduction.

Salty and sour taste appear to use the simplest signal transduction mechanism: ionic transport. Salty taste involves direct depolarization through Na^+ channels in the microvilli. Sour taste involves H^+ blockage of K^+ channels leading to depolarization.

Bitter taste utilizes gustducin G_g (a homologue of retinal transducin) in signal transduction and follows a similar pattern. The steps include activation of the $G_{g\alpha}$ subunit, activation of cyclic GMP phosphodiesterase, decrease in cGMP, closure of Na^+ channels, and hyperpolarization of the taste cell leading to bitter taste. Sweet taste functions through G_s, ultimately resulting in depolarization. The mechanism includes activation of the $G_{s\alpha}$ subunit, elevation of cAMP levels, and closure of K^+ channels. Signal transduction is dealt with in more detail in Chapters 9 and 27.

LAYERS OF THE DIGESTIVE TUBE

Most students make every effort not to learn the layers of the digestive tract. This avoidance behavior is not beneficial and makes it incredibly difficult to distinguish the organs of the digestive system (Chap. 28).

The four layers of the wall of the digestive tract from inside to outside are mucosa, submucosa, muscularis externa, and serosa (or adventitia).

The mucosa consists of three individual layers: (1) epithelium, (2) lamina propria (the loose connective tissue underlying the surface epithelium), and (3) a thin muscularis mucosa.

The epithelium of the GI tract varies from a stratified squamous epithelium at the two ends to a simple columnar type in between, which is modified to form glands. The muscularis mucosa is always composed of smooth muscle.

The submucosa is dense irregular connective tissue found between the muscle of the mucosa and the thicker external muscle layers. It is vascular and contains a neural plexus (Meissner's or submucosal plexus).

The muscularis externa consists of an inner circular and an outer longitudinal layer with a ganglionic (Auerbach's or myenteric) plexus between the two layers.

The muscularis externa regulates the size of the lumen (inner circular layer) and the rhythmic movements of the GI tract, called peristalsis (outer longitudinal layer).

The serosa or adventitia (where covered by a simple squamous epithelial lining or mesentery) is the outermost layer of the GI tract. It consists of dense irregular connective tissue and contains nerves and blood vessels.

ESOPHAGUS

The layers of the GI tract are most obvious in the esophagus. However, the esophagus is not just a tube or passageway from the pharynx to the stomach. It does serve several important functions: (1) digestion which is initiated in the mouth is continued in the esophagus, and (2) mucus is added to the food bolus to facilitate swallowing at the superior end and protection of the wall against reflux of gastric acid at the inferior end of the esophagus. The muscularis externa undergoes a transition from skeletal muscle in its upper third to a mixture of skeletal and smooth muscle in the middle third. The lower third of the esophagus contains all smooth muscle.

STOMACH

The stomach (Fig. 21-2) consists of four regions: (1) cardia, (2) fundus, (3) body, and (4) pylorus. The esophagogastric (cardiac) sphincter guards the entrance to the stomach, whereas the stronger pyloric sphincter regulates outflow to the small intestine.

The cardiac portion of the stomach contains mostly mucous glands (see table). The fundus and body contain the gastric glands with the majority of cell types involved in gastric function. The pylorus contains predominantly mucous glands with hormone-secreting enteroendocrine cells interspersed.

The parietal cell is the unique cell of the gastric gland.

The parietal cell, like the osteoclast of bone, generates H^+ from carbonic acid.

The *parietal cell* uses a system similar to the osteoclast (Chap. 17). *Carbonic anhydrase* catalyzes the formation of carbonic acid, which dissociates into H^+ and HCO_3^-. The *bicarbonate* concentration is at least partially responsible for

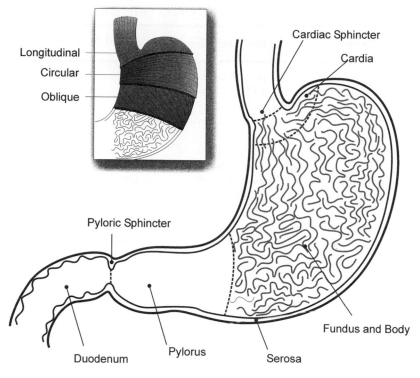

Figure 21-2
Regions of the stomach and the unique arrangement of the muscle layers to facilitate the generation of the chyme.

the neutrality of the parietal cell cytoplasmic pH. It is also released into the blood-stream following a meal, resulting in the postprandial (posteating) elevation of blood pH. Cl^- as well as H^+ is actively transported into the canaliculi (see below) to eventually reach the lumen.

The ultrastructure (electron microscopic view) of the parietal cell is shown in Fig. 21-3. The cell contains prominent *intracellular canaliculi* lined by *microvilli*. Another hallmark of the parietal cell is the presence of tubulovesicle pools. Parietal cell membrane shuttling and recycling function in an analogous fashion to the membrane handling in exocytosis-endocytosis. Following secreta-gogue stimulation of a parietal cell, *tubulovesicles* undergo exocytosis into the intracellular canaliculi and the microvilli (Fig. 21-3B). When the cell returns to the resting state, the excess membrane is endocytosed and stored in the tubulovesicle pool (Fig. 21-3A).

EPITHELIAL CELLS IN THE GASTRIC GLANDS

CELL TYPE	STRUCTURE	FUNCTION
Mucous (surface and neck) cell	Surface cells form the lumenal lining and dip into pits; neck cells are found in the narrowing of the gland	Produces mucins
Parietal cell	Large, pyramidal shape, eosinophilic staining	Produces HCl and gastric intrinsic factor
Chief cell	Triangular cell with zymogen granules	Secretes pepsinogen (active form is pepsin that hydrolyzes proteins to form peptides)
Enteroendocrine cell	Small cells with secretory vesicles polarized toward basal surface in close proximity to blood vessels	Produces peptide hormones and serotonin

The parietal cell possesses receptors for histamine, acetylcholine, and gastrin on its basal surface. Binding of secretagogue to receptor leads to the transport of membrane to form microvilli and the production of acid.

The parietal cell also produces gastric intrinsic factor, which is essential for the absorption of vitamin B_{12} from the ileum.

Vitamin B_{12} binds to gastric intrinsic factor and is required for normal erythrocyte development. In its absence, the result is pernicious anemia, which appears to be an autoimmune disease. (See Chap. 19.)

SMALL INTESTINE

The small intestine consists of three segments: *duodenum*, *jejunum*, and *ileum*. The duodenum receives the pancreatic duct and the common bile duct carrying the pancreatic juice and bile respectively.

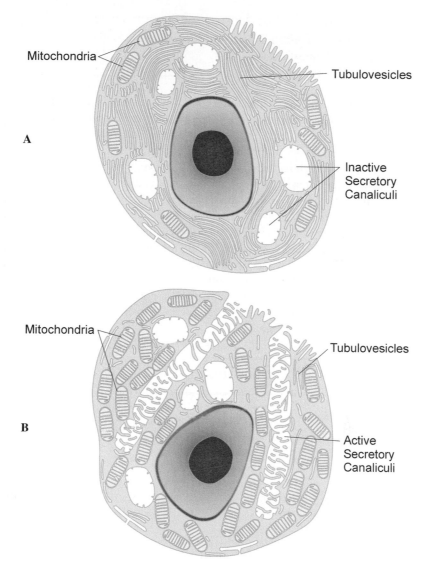

Figure 21-3

A. Ultrastructure of resting parietal cell. The extensive tubulovesicles and small, inactive secretory canaliculi are the distinctive features. *B.* Ultrastructure of a stimulated parietal cell. The tubulovesicular membranes have been exocytosed to the secretory canaliculi.

> The small intestinal epithelium maximizes surface area at the macroscopic, microscopic, and ultrastructural levels.

The structure of the small intestine is directly related to its absorptive function. There are three orders of folding. From a teleologic perspective, the increase in surface area is present to facilitate the absorptive digestive function of the small bowel. The first order of folding (macroscopic level) is the presence of *plicae*, folds of mucosa each including a submucosal core. The next level of folding and increased surface area (microscopic level) are the *crypt-villus* systems. Each fingerlike villus contains a core of lamina propria. The next order of increased surface area (ultrastructural or EM level) is made up of the *microvilli*. These structures are apical specializations of the columnar cells and are covered with a glycoprotein and glycolipid coat called the glycocalyx. The *glycocalyx* contains an array of brush border enzymes (disaccharidases: maltase, sucrase, isomaltase, lactase, small amounts of lipase, and peptidases) produced by the small intestinal epithelial cells. The microvilli are supported by a core of *actin* microfilaments that provide support for the villi that undergo rhythmic contractions.

EPITHELIAL CELLS IN THE CRYPTS OF LIEBERKUHN

CELL TYPE	STRUCTURE	FUNCTION
Goblet cell	Shape of goblet	Produces mucins
Enterocyte (small intestinal absorptive cell)	Columnar epithelial cell, basal nucleus, perinuclear Golgi, apical brush border containing disaccharidases	Absorption of protein, carbohydrate, and lipid
Paneth cell	Basal crypt cell with prominent, eosinophilic granules	Produces lysozyme that digests bacterial cell walls and controls intestinal flora
Enteroendocrine cell	Chromaffin (stain with chromium), argyrophilic (stain with silver stains)	Produces peptide hormones and serotonin

The digestion and subsequent absorption, which occurs in the small intestine, require the addition of enzymes and bile from the pancreas and the gallbladder respectively. The *crypts and villi* constitute the intestinal glands of Lieberkühn that release mucus and other secretions into the lumen of the small intestine (see table and Fig. 21-5). The *brush border enzymes* present on the apical surfaces of the enterocytes play a key role in carbohydrate digestion. In addition, *enteropeptidase (formerly known as enterokinase)* is essential for the activation of zymogens and proenzymes in the pancreatic juice. In addition, there are the *submucosal glands of Brunner* in the duodenum that are responsible for the formation of bicarbonate and mucus. These secretions counteract the acidity of the stomach and produce the neutral to alkaline pH optimal for action of the pancreatic juice and bile. Additional bicarbonate is added in the pancreatic juice.

DIGESTION

The small intestinal enterocyte absorbs monoglycerides and glycerol from fat digestion, peptides from protein digestion, and monosaccharides from carbohydrate breakdown.

The brush border of the enterocyte contains disaccharidases for carbohydrate digestion. Amino acids and monosaccharides are actively transported by specific carriers, and fatty acids and glycerol are passively transported into the enterocyte.

Digestion of carbohydrates

Digestion of carbohydrates begins in the mouth with the addition of mucus and amylase that hydrolyzes starch to form sugars, specifically disaccharides such as maltose and oligosaccharides. This digestion continues in the esophagus and small intestine. In the small intestine, more amylase is added in the pancreatic juice. Finally, the disaccharidases of the brush border break down sugars to monosaccharides (i.e., glucose, galactose, and fructose) which are absorbed by the enterocytes by an active, energy-dependent, process involving mediator-specific carriers.

There are several clinical issues regarding carbohydrate digestion. The first of these is lactase deficiency. The insufficiency of lactase, required for digestion of lactose, the major carbohydrate moiety found in milk, is a relatively common medical problem. Individuals with this deficiency have difficulty digesting milk and dairy products. The second is important in terms of the move toward "high-fiber" diets. The increase in abdominal gas (CH_4, H_2, and CO_2) and flatulence after intake of large quantities of beans and other legumes is due to the presence of carbohydrate bonds that can be broken down only by bacterial enzymes. The bacterial activity occurs in the distal small bowel (ileum) as well as in the colon. This hydrolysis of carbohydrate bonds generates the above-mentioned gases and other substances that stimulate intestinal motility.

Digestion of protein

Digestion of protein begins with the hydrolysis of proteins to peptides in the stomach. This breakdown occurs initially through the action of pepsin that is activated from pepsinogen in the acidic milieu of the stomach. Breakdown of polypeptides also occurs in the small intestine through the action of trypsin, chymotrypsin, elastase, and carboxypeptidases. These pancreatic enzymes are secreted into the pancreatic juice as the zymogens *trypsinogen* and *chymotrypsinogen*, respectively. *Carboxypeptidases* are secreted into the pancreatic juice as the proenzymes *procarboxypeptidases*. The activation cascade for these proteolytic enzymes and peptidases is initiated by enteropeptidase (enterokinase) secreted by duodenal enterocytes. Further digestion of the resulting smaller peptides (oligopeptides) occurs through the action of peptidases in the small intestinal brush border. The amino acids which result from protein digestion are actively transported into the enterocyte by a carrier-mediated process.

Digestion of fat

There are a number of factors that function together to facilitate the digestion of fat.

Digestion of lipids is problematic because of their lack of solubility. The small intestine "solves" this problem by (1) emulsification, formation of small lipid droplets in an aqueous lumenal fluid sea, and (2) solubilization of lipids to form micelles through the action of bile acids (salts).

Lipid digestion (Fig. 21-4) begins with lipid entering the small intestine primarily in the form of triglyceride (triacylglycerol) droplets. *Lipase* from the pancreatic juice splits these large droplets into smaller emulsion droplets and releases fatty acids and glycerol that carry out a surfactant function. The bile acids form aggregates with a regular orientation (hydrophobic regions toward the inside and hydrophilic regions toward the outside, in contact with the lumenal fluid). These micelles interact with the lipids to form an intestinal mixed micelle. The very hydrophobic regions of the lipids are most protected on the inside of this micelle. The micelles provide the mechanism for transport of fatty acids, monoglycerides (monoacylglycerols), glycerol, phospholipids, cholesterol, and vitamins A and K.

Fats must be broken down to the level of monoglycerides and glycerol in order to be absorbed by enterocytes in the small bowel.

Lipases in the pancreatic juice hydrolyze neutral fats to fatty acids and glycerol. Cholesterol esters are broken down by *lipid esterase*, which splits cholesterol esters into cholesterol and fatty acids. Micelles of monoglycerides and

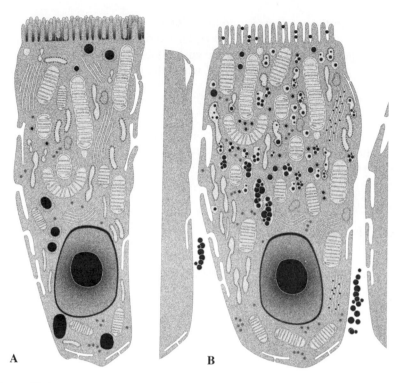

Figure 21-4
A. A small intestinal enterocyte at the electron microscopic level in the resting state. Note the presence of extensive RER. Why is there so much RER in a cell dedicated to absorption? (The enterocyte synthesizes brush border enzymes and the protein portion of chylomicra.) *B.* A small intestinal enterocyte at the electron microscopic level postprandial, following a meal containing lipid. Note the release of the chylomicra into the lacteals at the lateral borders of the enterocyte.

glycerol are passively absorbed by diffusion through the enterocyte plasma membrane.

Synthesis of protein in the RER and triglycerides in the SER are critical to the absorption of lipid breakdown products.

The monoglycerides and glycerol are reesterified in the smooth endoplasmic reticulum (SER) of the enterocyte to form trigylcerides. The triglycerides are recombined with proteins to form the *chylomicra*. These lipoprotein molecules are released from the enterocytes into the neighboring *lacteals* (Fig. 21-4). The lacteals are lymphatic vessels that begin as dead-end lymphatic capillaries in the villi, form larger lymphatic vessels in the gut wall, and subsequently empty into an abdominal reservoir, the cisterna chyli, which drains via the thoracic duct into the venous system.

In contrast to lipid absorption that is transmitted through the lacteals, carbohydrate and protein products are transported to the liver by way of the hepatic portal system.

LARGE INTESTINE

The large intestine consists of the cecum and appendix; ascending, transverse, and descending colon; and rectum. The epithelium of the large bowel is a simple columnar epithelium in which there are crypts, but *no* villi.

The large intestine resorbs water and provides lubrication.

The functions of the large intestine include: (1) secretion of mucus for lubrication (the goblet cell is the predominant cell of the large intestine), (2) absorption of fluid, (3) formation of the fecal mass, and (4) continuation of digestion initiated in the small intestine and that added by putrefying bacteria. The appendix is primarily an organ of the immune system in humans in which there is little cellulose breakdown. The appendix is in excellent position to sample the microbial flora of the large bowel. The rectum leads to the anus where the epithelium returns to a stratified squamous type and skeletal muscle returns as the external sphincter.

CELL TURNOVER

Rapid cell turnover is a protective mechanism used by the GI epithelium.

Most of the GI tract possesses an epithelial lining that turns over rapidly. This is particularly true in the stomach and small intestinal epithelia where the lining is replaced about every 5 days in humans. In the esophagus and anus, new cells are formed in the basal layer of the stratified squamous epithelium by cell division and migrate through the suprabasal levels to be sloughed off into the lumen. In the small intestine (Fig. 21-5) new cells are formed in the crypts from undifferentiated cells (*proliferative compartment*).

The unitarian stem cell exists in the small intestinal crypt.

All four cell types, enterocytes, goblet cells, enteroendocrine cells, and Paneth cells, are derived from stem cells located in the crypts. New cells differentiate and migrate into the *differentiating compartment* on the villus. In the large intestine, where there are no villi, the *proliferative compartment* is found at the base of the crypts. Newly formed cells migrate into the differentiation compartment in the upper part of the crypt (gland). In the gastric glands of the stomach, the stem cells are found in the neck of the glands. These stem cells differentiate into all four types, mucous (surface and neck), enteroendocrine, parietal, and chief cells. The surface and neck cells turn over most rapidly, whereas the other cells are replaced more slowly.

ENTEROENDOCRINE CELLS

These cells function as unicellular glands forming peptide hormones and are analogous to the unicellular goblet cells that produce mucus. The enteroendocrine cells are formed from endodermal stem cells in all regions of the GI tract. Enteroendocrine cells secrete their hormones in a number of different ways:

1.   Endocrine (into the bloodstream)
2.   Paracrine (adjacent cells are the target)
3.   Autocrine (The secreting cell is also the target for the hormone; see Chap. 22.)

GUT-ASSOCIATED LYMPHOID TISSUE (GALT)

The bulk of the body's immune defenses is centered in the GALT.

GALT includes transitory aggregations of lymphocytes, neutrophils, and eosinophils as well as permanent structures. The permanent structures include the

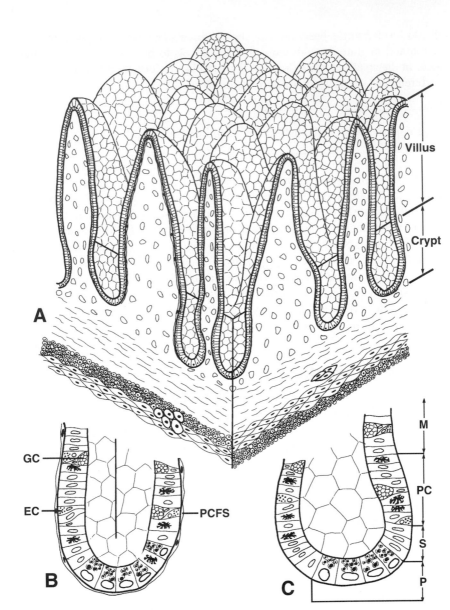

Figure 21-5
All the cell types of the small intestinal epithelium are derived from a single stem cell (unitarian stem cell theory). The enterocytes (small intestinal absorptive cells), Paneth cells (PC), goblet cells (GC), and enteroendocrine cells (EC) are all derived from stem cells located in the crypts of the small intestine. The stem cells (S) are located within the progenitor zone (P) of the crypt. The pericryptal fibroblast sheath (PCFS) is a layer of fibroblasts located beneath the epithelium in the underlying lamina propria. (Reproduced with permission from Cormack DH: *Ham's Histology*, 9th ed. Philadelphia: Lippincott, 1987.)

appendix, Peyer's patches, and mesenteric lymph nodes. The appendix and *Peyer's patches* are in position to sample antigens from the gut lumen. Peyer's patches are dome-shaped lymphoid structures under the mucosal surface. Peyer's patches contain B- and T-cell–dependent areas like other secondary lymphoid organs and possess *high endothelial venules (HEVs)*, which facilitate entrance of lymphocytes into lymphoid organs from the bloodstream. The mucosal covering of the dome of the Peyer's patches includes specialized M (microfold) cells, which sample particulate antigen and present it to antigen-presenting cells in the underlying lamina propria. The antigen-presenting cells (APCs) and macrophages phagocytose the antigen and present it to helper and cytotoxic T cells as well as B lymphocytes. (See Chap 19.)

Both soluble and particulate antigens are presented to APCs (macrophages) in the lamina propria of the gut.

Soluble antigen is picked up by the small intestinal enterocytes, is phagocytosed by macrophages and presented to lymphocytes in the lamina propria in the core of the villus. The lymphocytes are activated in the lamina propria, but travel to the mesenteric lymph nodes for subsequent maturation. Following maturation that includes class switching and deletion of inefficient clones (see Chap. 19), cells that are exposed to antigen in the gut wall develop appropriate surface markers. These cell surface molecules recognize *addressins* on the HEVs of mesenteric lymph nodes or Peyer's patches and specifically target lymphocytes.

GALT makes optimal use of antigen surveillance.

GALT depends on *lymphocyte recirculation* in which memory T and B cells provide a surveillance system for antigen entering the GI tube. An additional part of the surveillance system is the intraepithelial lymphocyte, found in between epithelial cells. Intraepithelial lymphocytes are primarily CD8+ cells prepared to attack virus-infected cells. They have a different phenotype of surface markers from the lymphocytes in the circulating pool.

The primary immunoglobulin produced by GALT is IgA (secretory IgA, sIgA).

IgA, also known as *secretory immunoglobulin (sIgA)*, is synthesized and secreted by plasma cells in the lamina propria of the gut. The sIgA is picked up at the basal surface of enterocytes and transported across the cell. While in the cell, the sIgA is linked to a protein called *secretory component* that inhibits degradation of sIgA by proteolytic enzymes in the GI lumen. Secretory IgA, unlike IgG, does not stimulate the complement system, but functions by coating microorganisms, thus inhibiting microorganism binding to the epithelium.

GLANDS

·

- **Exocrine Glands: Salivary Glands**
- **Pancreas**
- **Liver**
- **Gallbladder**
- **Cystic Fibrosis**
- **Endocrine Glands**
- **Pituitary**
- **Thyroid**
- **Suprarenal (Adrenal) Glands**
- **Parathyroid Glands**
- **Pineal Gland**
- **Other Glands**

· · · · · · · · · · · ·

The epithclia of the body in different systems are modified for secretory function.

> The glands are subdivided into two main types: (1) exocrine glands, in which the secretory product is carried in a duct system to the lumen or the surface of the organ, and (2) endocrine glands, which secrete their product(s) directly into the blood in the absence of ducts.

The exocrine glands include the major and minor salivary glands that produce saliva. The mucus and enzymes that make up the saliva empty into the oral cavity by way of ducts. Endocrine glands include the pituitary, thyroid, pineal, parathyroids, adrenals, and male and female gonads. Other glands in the body such as the pancreas and liver exhibit both endocrine and exocrine secretory characteristics.

> The simplest type of gland is the unicellular gland that may function in an exocrine or an endocrine manner.

The goblet cell of the small intestinal epithelium is an example of a unicellular exocrine gland. The goblet cell releases mucins into the crypt, which functions as a duct. The enteroendocrine cells are unicellular endocrine glands which release their secretion, usually from the basal surface, to reach blood vessels traversing nearby. In addition to these traditional classifications of secretion, there are two more mechanisms of cellular secretion. *Paracrine* secretion affects neighboring cells, whereas *autocrine* secretion is cellular self-stimulation; the same cell is the source and target for the secretory product. Many growth factors function in a paracrine manner influencing adjacent cells. IL-2 secretion by activated T cells is an example of autocrine function. The activated T cells also possess IL-2 receptors, which allow the T cells to maintain their own activated state.

Glandular secretion is often equated with the multicellular glands. The exocrine glands are classified on the basis of the presence or absence of ductal branching (compound versus simple without branching) and the type of secretory ending [*tubular* (elongate) or grape-like clusters (*acinar*)]. The secretory component plus the duct system is known as the *parenchyma* of the gland; the surrounding connective tissue and capsule form the structural support known as the *stroma*. The parenchyma of endocrine glands consists of clusters of cells surrounding sinusoids or follicles surrounded by capillaries. Some of the major glands such as the liver and pancreas have both endocrine and exocrine function.

> There are a number of important classifications of the exocrine glands—holocrine, apocrine, and merocrine—in which the whole cell, the apical cytoplasm, or no part of the cell is lost during secretion.

Holocrine glands are just what they sound like, the "whole" cell dies as the entire cell is released with the secretory product. Sebaceous glands are the most prominent example of a holocrine gland. *Apocrine glands* also are what they sound like; they release their *apical* cytoplasm along with their secretory product. Sweat glands located in the axillary region are examples of apocrine glands. In *merocrine glands*, cytoplasm is retrieved in a conservation process with no loss of membrane or cytoplasm. This process, known as exocytosis, occurs where secretory product is stored in vesicles which fuse with the apical membrane while the secretion is released to the external environment. (See Chap. 6.) Most glands are merocrine, and the pancreas and salivary glands are two examples of this manner of secretion. The mammary gland represents a gland which has attributes of both merocrine (milk proteins) and apocrine (lipid) secretion (Chap. 25).

EXOCRINE GLANDS: SALIVARY GLANDS
Minor salivary glands

There are minor salivary glands throughout the oral cavity, including the tongue (von Ebner's), lips (labial), palate (palatal), and cheeks (buccal). These glands secrete continuously and keep the mucosa of the oral cavity moistened.

Major salivary glands

There are three major salivary glands, the *parotid* (located between the mandible and the mastoid process, just under the pinna of the ear), the *submandibular* (located inferior to the body of the mandible), and the *sublingual* (located in the floor, or "gutter" of the mouth). Each of the glands is organized in the same way. Think of grapes on a grapevine (Fig. 22-1), where the grapes represent acini composed of acinar cells surrounded by a myoepithelial cell (cell of epithelial origin with extensive actin bundles allowing contractility) which squeezes secretion from the acinus into the duct system. The vine represents the duct system. The grapes and vines are partitioned by connective tissue into lobules so that some of the duct system is within each lobule (*intralobular ducts*) and some between lobules (*interlobular ducts*). The intralobular ducts are of two types. The intercalated duct drains the acinus, consists of squamous to low cuboidal cells, and appears to be involved in HCO_3^-/Cl^- exchange: HCO_3^- secretion/Cl^- reabsorption.

The striated ducts are critical in active transport and the conversion of the saliva from an isotonic (primary secretion) to the hypotonic state. In the striated ducts, Na^+ is actively absorbed from the saliva, Cl^- is passively absorbed from the saliva, and K^+ is actively secreted into the saliva.

In the striated ducts, the Na^+ concentration is greatly reduced, and the K^+ concentration is increased. K^+ is exchanged for Na^+, but at a slower rate so that a negative voltage is created across the duct leading to more passive Cl^- movement. The striated ducts are composed of columnar cells with basal striations, which represent extensive accumulations of mitochondria for the active transport that occurs in these ducts. The interlobular ducts conjoin to form the duct system (vine) that eventually opens into the oral cavity.

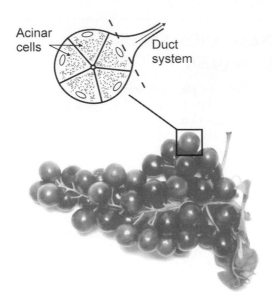

Figure 22-1 The structure of an acinar-exocrine gland.
Photograph of grapes on a vine. This arrangement is analogous to the acinar structure of
glands found in the salivary glands and pancreas. The grapes represent acini, each composed of five to seven individual acinar cells (see inset). The secretion of the acinar cells
is released into the duct structure (represented by the vine). The vine extends from the
acini to the oral cavity where the salivary gland secretion is released into the mouth to
faciliate digestion. The connective tissue of the gland capsule and septa surrounds and
divides the gland into separate lobules. Myoepithelial cells surround the acini and squeeze
the secretion out of the acinus following neural stimulation.

The major salivary glands contain mucous, serous, and mixed acini, which
produce mucus and amylase as well as the ionic and fluid components of the
saliva.

The salivary glands are classified as *mucous* (thick mucin-containing secretion), *serous* (watery, enzyme-filled secretion), or *mixed* (mucous and serous
secretion) types. The parotid gland is a serous gland, whereas the submandibular
and sublingual glands are mixed glands. (The submandibular gland is predominantly serous; the sublingual gland mostly mucous.) Acini may also be serous,
mucous, or mixed. The mixed type consists of mucous acini with a cap of serous
cells called a *serous demilune*. In the case of the salivary glands the serous secretion contains primarily the enzyme *amylase*.

The salivary glands are under the direct control of the autonomic nervous system, and the parasympathetic nerves are the primary regulators of gland function.

The autonomic nerves originate in the salivatory nuclei in the brainstem and eventually travel with peripheral nerves to reach the parasympathetic ganglia embedded in the interlobular connective tissue of the salivary glands. Parasympathetic secretion produces an enzymatic (serous) secretion. Sympathetic fibers travel with the blood vessels. Stimulation of the sympathetic nerves results in a thick, mucous secretion.

PANCREAS

The pancreas is a both an exocrine gland producing the pancreatic juice and an endocrine gland (islets of Langerhans) producing insulin, glucagon, and somatostatin.

The pancreas is located in the abdomen with its head sitting in the C shape of the duodenum and its tail touching the spleen. It produces the pancreatic juice synthesized by pancreatic acinar cells. The pancreatic juice contains *amylase*, *RNAse*, *DNAse*, *cholesterol esterase*, *carboxypeptidase*, *trypsinogen*, *chymotrypsinogen*, and *lipase*. The proteolytic enzymes are secreted as proenzymes, which are coverted to their active form by *enterokinase* in the small intestine. Trypsin can also activate trypsinogen. Trypsin inhibitor is synthesized and stored with the proenzymes in the pancreas; it inhibits premature activation, which could lead to *acute pancreatitis*.

The exocrine secretion of the pancreas is regulated by parasympathetic fibers from the vagus (tenth cranial nerve) and two hormones, cholecystokinin (CCK, pancreozymin) and secretin. The primary regulator is CCK, which stimulates enzyme secretion from acinar cells, whereas secretin stimulates a bicarbonate-rich fluid, ductal secretion.

Unlike the salivary glands which are regulated by the autonomic nervous system, CCK primarily controls the exocrine pancreas. Enteroendocrine cells in the pylorus and duodenum produce cholecystokinin and secretin. These two hormones travel by way of the bloodstream to the pancreas.

The islets of Langerhans are the endocrine portion of the pancreas and secrete glucagon (α-cells), insulin (β-cells), somatostatin (δ-cells), vasoactive intestinal peptide (D_1 cells), and pancreatic polypeptide (PP cells).

The endocrine secretion of the pancreas is regulated primarily by blood glucose levels, but there are feedback loops which complicate pancreatic endocrine secretion. Elevated blood sugar levels stimulate insulin secretion. Other factors which stimulate insulin secretion include elevated amino acid concentration, GI hormones (gastrin, secretin, cholecystokinin), and other hormones (growth hormone, glucagon, pancreatic polypeptide, vasoactive intestinal peptide, and others). Some of these hormones are localized within specific cell types that may be identified by immunocytochemistry with specific antibodies.

Insulin is the hypoglycemic factor. It reduces elevated blood sugar levels with effects on the liver, muscle, fat, and protein.

Insulin promotes liver uptake, storage, and use of glucose and its conversion into fatty acids. Insulin promotes the *transport of glucose* into muscle and the *storage of glycogen* in muscle. Insulin increases glucose transport into most cells of the body except brain cells. This is particularly important in diabetic hypoglycemic shock where a protective coma is produced to "save" brain cells which are permeable to glucose at all times. In other cells, insulin stimulates uptake of glucose through upregulation of glucose transporters on the target cell surface. Insulin also enhances fat and protein storage.

Diabetes mellitus is a disease in which the beta (β) cells of the pancreas are deficient in their secretion of insulin. There are two types, type I insulin dependent (juvenile onset) and type II insulin independent (adult onset). The vast array of insulin effects on body metabolism is part of the reason for the extensive complications which occur in diabetic patients, particularly with poor blood sugar control.

Glucagon is the hyperglycemic factor that opposes the action of insulin. It is secreted in response to low blood sugar levels.

Glucagon stimulates the breakdown of liver glycogen (glycogenolysis) and increased gluconeogenesis. Glucagon secretion is regulated by blood sugar, blood amino acid levels, and exercise as well as other factors such as GI and pancreatic hormones.

LIVER

Like the pancreas, the liver is both an endocrine and exocrine organ.

Bile is the *exocrine* secretory product of the liver. *Endocrine secretion* is less tangible, but is usually characterized by glucose and lipoprotein, as well as plasma proteins (angiotensinogen and globulins except for the immunoglobulins produced by plasma cells).

The liver is unique among glands in that there appears to be no division of labor; all liver (hepatic) cells (hepatocytes) have the structural machinery and the capacity to carry out the complete range of functions of the liver.

The hepatocyte is an ideal cell to view all the typical cytoplasmic organelles at the electron microscopic level. The hepatocyte has elaborate protein synthetic machinery, cholesterol processing organelles, active smooth endoplasmic reticulum, and both apical (microvilli) and basolateral (canaliculi) specializations.

The liver is the heaviest gland in the body and is essential for life. It receives a rich blood supply from the hepatic artery (carrying oxygenated blood from the aorta) and the hepatic portal vein (carrying blood from the GI tract rich in nutrients). The liver produces bile, which is transported through a duct system to the gallbladder.

The histology of the liver can be described as rows of interlocking plates and cords of cells anastomosing around sinusoids. Interspersed around the liver parenchyma are *portal triads* containing the three main components of the hepatic vascular and bile duct system. At other sites, prominent veins are found within the parenchyma for drainage of blood from the liver. There is little stroma except for the thin (Glisson's) capsule and sparse connective tissue interspersed between the liver parenchyma.

Blood and bile flow

Blood flow and bile flow occur in opposite directions.

This is an important concept when considering the three different views of hepatic lobulation and the structure-function relationships of the liver discussed in Fig. 22-2. Blood from the hepatic portal vein and hepatic artery in the portal triads travels through the sinusoids to the central veins. Bile flow is from the hepatocytes toward the bile ducts in the portal triads.

Hepatic lobulation

Lobulation is not well defined in the human compared with the pig liver. As a result there is a mismatch between a functional lobule and a morphologic unit of the gland.

The three classifications of liver lobulation are shown in Fig. 22-2 and are described below. In the pig liver, there are regular hexagonal arrays of portal areas surrounding a central vein. This is known as a *classic lobule* and is difficult to visualize in the human liver. Since it is based on functional flow from the portal areas toward a central vein, it emphasizes the endocrine function of the liver. The next functional unit of the liver is the *portal lobule*, which has three central veins surrounding a central portal triad. It focuses on flow of bile toward the bile ducts in the portal areas, which emphasizes the exocrine function of the liver. There is a third way of viewing the functional lobulation of the liver. The *liver acinus* emphasizes the actual or functional hepatic blood supply. The hepatic portal veins and hepatic arteries are not terminal vessels that empty directly into sinusoids. These vessels progressively decrease in size as they extend and branch in all three axes from the portal triad descending to the diameter of a sinusoid. The acinus directly relates blood flow to hepatic metabolism and is reflective of repair and regenerative processes. Areas with richer blood supply regenerate first. The core of the acinus thus receives a blood supply richer in oxygen and nutrients. In "fat" times, hepatocytes in the core (region one) surrounding the portal and arterial supply are the first to lay down glycogen deposits. Conversely, they are also the first to give them up (glycogenolysis) in lean times.

• **HEPATOCYTE**   The primary cell in the liver is the hepatocyte. There is polarity of hepatocytes. The apical surface along the sinusoid possesses numerous microvilli. These microvilli extend into the perisinusoidal *space of Dissé*, which allows blood to come close to the hepatocyte and facilitates absorption. Along lateral surfaces the bile canaliculi form from the membranes of adjacent hepatocytes. The bile canaliculi are a lateral domain with apical domain characteristics. They are sealed off from the rest of the cell by tight junctions, the Golgi are "aimed" at the bile canaliculus, and the bile is secreted into the lumen of the canaliculus.

• **KUPFFER CELLS**

The macrophages and antigen-presenting cells of the liver are derived from monocytes at some time during development and are known as Kupffer cells.

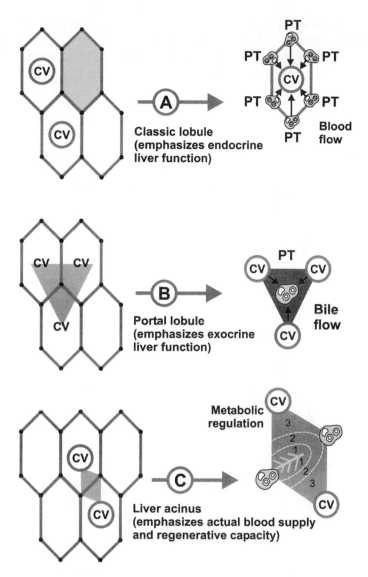

Figure 22-2 Liver lobulation.
The three classifications of liver lobulation. The classic lobule emphasizes the endocrine function of the liver. The portal lobule emphasizes the exocrine function of the liver, and the liver acinus represents true "functional" blood supply and metabolic regulation of the liver.

Kupffer cells reside in the hepatic sinusoids and phagocytose debris and dying RBCs. They also secrete and respond to cytokines in a similar fashion to other macrophages in the body. A small portion of the bile pigment is derived from the breakdown of hemoglobin by Kupffer cells.

• **OTHER CELLS OF THE SINUSOIDS**   The sinusoids begin at the periphery of the lobule from branches of the portal vein and hepatic artery. They are lined by endothelial cells and mesenchymally derived fat-storing fibroblasts (cells of Ito for lovers of eponyms). The fibroblasts appear to have the capacity to store vitamin A. The sinusoids are incomplete with gaps and fenestrations. The incomplete basal lamina results in the absence of a barrier between the sinusoids and the sinusoidal space. The result of this barrier-free situation is that plasma can come in direct contact with hepatocytes, facilitating transport bidirectionally with the plasma.

• **LIVER FUNCTION**

FUNCTIONS OF THE LIVER

Maintenance of blood glucose levels
Secretion of lipoprotein
Secretion of bile
Storage of vitamins A and B
Secretion of plasma proteins: albumin, globulins, and fibrinogen
Synthesis of cholesterol
Phagocytosis of old erythrocytes
Detoxification of toxic materials in the smooth endoplasmic reticulum
Hematopoiesis in the fetus

Hepatocytes carry out the functions listed in the table. The hepatocyte, along with steroid-hormone producing cells, is one of the few cell types with a prominent system of smooth endoplasmic reticulum (SER).

Many hepatocyte functions are closely associated with the SER.

The hepatocyte, and specifically its SER, is directly involved in glycogen metabolism (See Fig. 22-3). The hepatocyte, under the influence of insulin, removes glucose from the blood and stores it as glycogen (glycogenesis) in the cytoplasm. This accounts for the accumulation of glycogen seen in the hepatocyte at the EM level in close association with the SER. Glycogen granules consist of glucose polymers (glycogen) enveloped by the enzymes responsible for formation (glycogen synthase) and breakdown (glycogen phosphorylase) of glycogen.

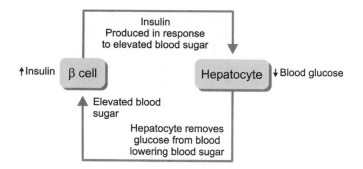

Figure 22-3 Negative feedback regulation of hormonal secretion.
The diagram illustrates negative feedback as occurs between the islets of Langerhans and hepatocytes in the liver. Elevated blood sugar, following a meal, results in stimulation of the pancreatic β cell to release stored insulin and synthesize more insulin for secretion. The result is elevated insulin levels in the bloodstream. The elevated serum insulin stimulates the hepatocyte to remove glucose from the bloodstream and store it as glycogen. The result is a decreased blood sugar level which negatively feeds back on the β cell to shut off insulin secretion.

The SER contains glucose-6-phosphatase, the enzyme responsible for conversion of glucose-6-phosphate into glucose. In response to glucagon, the hepatocyte produces more SER membrane for glycogen breakdown to glucose (glycogenolytic activity) and for lipid synthesis.

The SER of hepatocytes proliferates in the presence of drugs (e.g., barbiturates) and toxic compounds and is involved in the detoxification process. The SER contains enzymes required for oxidation, methylation, and conjugation necessary for drug detoxification. The cytochrome P_{450} family of enzymes is responsible for the conversion (hydroxylation) of water-insoluble drugs (hydrocarbons) to a water-soluble form so that they may be excreted in the urine. The synthesis of cholesterol involves the SER as does the lipid portion of hepatic lipoproteins. The hepatocyte rough endoplasmic reticulum (RER) synthesizes protein that is added to lipid synthesized in the SER to form lipoprotein.

The hepatocyte secretes plasma proteins (globulins, except for the immunoglobulins) into the bloodstream and bile into the bile canaliculi. Kupffer cells are involved in the breakdown of old erythrocytes and recirculation of their pigment into bile. The liver functions as a *hematopoietic organ* in development, before the bone marrow becomes hematopoietic.

• **BILE FORMATION** The hepatocytes secrete bile continuously into the *bile canaliculi*, which ultimately lead to the extrahepatic ductal system. Bile is composed of an aqueous solution containing cholesterol, lecithin, bile acids, bile pigments (*bilirubin*) and electrolytes. Bilirubin is formed from the breakdown of hemoglobin by Kupffer cells in the liver, but predominantly by macrophages in the sinusoids of the spleen (see Chap. 19). Bilirubin is bound to albumin and circulates to the liver where hepatocytes internalize the bilirubin plus albumin complex and conjugate bilirubin to glucuronic acid through the action of *glucuronyl-transferase* located in the SER. The conjugated bilirubin is water-soluble and released into the bile canaliculus and eventually the duodenum. The bile acids are the other major component and are mostly reutilized from the distal intestine through a bacterial breakdown process coupled to intestinal epithelial absorption and return to the liver (enterohepatic recirculation). About 10 to 20 percent of the bile acids are synthesized by the smooth endoplasmic reticulum from cholic acid (a cholesterol derivative) and the amino acids glycine and taurine.

Jaundice is a yellowing of the skin due to the presence of bile pigments in the blood. It is a sign of liver disease or interference with the flow of bile through the duct system.

Jaundice may be due to hepatic failure or obstruction of the bile duct. Jaundice occurs in the week after birth in humans because of neonatal deficiency in *glucuronyltransferase*. In Craigler-Najjar syndrome there is a deficiency in this enzyme throughout life. Gilbert syndrome involves difficulties in hepatocyte uptake of bilirubin, and other syndromes involve abnormal bilirubin glucuronide excretion into the bile canaliculi. Obstruction of the bile duct by gallstones and pancreatic carcinoma that obstructs the entrance of the bile duct into the duodenum also cause jaundice.

• **REGENERATION**   Injury to the liver results in hypertrophy and hyperplasia (increase in hepatocyte size and number) to repair the damage. Repeated insults as occur in chronic alcohol consumption eventually lead to replacement of hepatocytes by fibrotic connective tissue, reduction of functional hepatocyte capacity, and expression of a panel of enzymes signaling the presence of cirrhosis of the liver.

GALLBLADDER

The gallbladder stores and concentrates the bile through absorption of water.

The gallbladder is stimulated to contract by CCK, the same hormone which stimulates production of the pancreatic juice. The gallbladder has a very thin wall and a simple columnar epithelium with folds that appear similar to shallow crypts that increase with age. Concentration of bile occurs through active Na^+ transport by the epithelium with water following the Na^+ osmotically. The gallbladder receives bile from the liver through the hepatic duct, which joins the cystic duct (carrying bile to and from the gallbladder) to become the common bile duct which enters the duodenum. Muscular sphincters also regulated by CCK control bile flow from the gallbladder.

CYSTIC FIBROSIS

Cystic fibrosis (CF) occurs with an incidence of 1 of 2000 caucasian births and is an autosomal recessive disorder. Organs affected include the respiratory and GI systems as well as the sweat glands. The disease is caused by a mutation in a gene on chromosome 7, called the cystic fibrosis transmembrane conductance regulator (CFTR).

Deletion of phenylalanine is the most common mutation in the CFTR gene. The CFTR gene encodes a single polypeptide chain, which functions as a cyclic adenosine monophosphate (AMP)-regulated Cl^- channel and as a regulator of other channels. The CFTR protein is a member of the ABC transporter superfamily named because of a common adenosine triphosphate (ATP)-binding cassette. The CFTR increases absorption and decreases secretion of electrolytes. The CFTR function may be different depending on the function of the epithelia. The small intestinal and airway epithelia are described as volume-absorbing, the sweat ducts are salt-absorbing, and the pancreas is volume-producing. In the airways, there is increased transport of Na^+ (water follows) and decreased CT permeability toward the serosa from the mucosa, resulting in dehydrated secretions and poor clearance. The CFTR fails to transport Cl^- into the lumen of the airway; intracellular Cl^- levels draw Na^+ and H_2O into the cell, resulting in thicker mucus. In the pancreas, impaired Na^+ and Cl^- transport into the ductal lumen results in retention of enzymes in the pancreas and eventual destruction of the pancreatic tissue (pancreatic insufficiency). In the intestinal epithelium, the lack of Cl^- and H_2O secretion coupled with increased water absorption results in an inability to clear mucins and other intestinal secretions. The result is dessicated intestinal contents and eventual obstruction of the intestine (meconium ileus). The bile ducts and gallbladder are unable to secrete water and salt, which also interferes with bile production and may lead to inflammation of the gallbladder. In the sweat glands, CF patients produce normal volumes of sweat, but the ducts are incapable of Cl^- absorption resulting in salty sweat. The salty sweat of these CF infants is the means for diagnosis and the basis of the quotation from northern European folklore: "Woe to the child who when kissed on the forehead tastes salty for he is bewitched and must soon die." Treatments with DNAse I to depolymerize the high concentration of DNA in the pulmonary sputum, coupling of a Na^+ channel blocker with drugs to stimulate non-CFTR-mediated Cl^- transport, and pancreatic enzyme replacement have been used to extend the lifespan of children diagnosied with CF. However, ultimately gene therapy may be the only way to change the outcome suggested in the folklore quotation. Adenovirus- and liposome-mediated delivery of CFTR to the epithelium of the airway is currently being attempted.

ENDOCRINE GLANDS

The endocrine system and the nervous system are the major integrative systems of the body.

Whereas the nervous system responds rapidly with a response of short duration, the endocrine system responds slowly and the response is of longer duration.

Endocrine glands function by a regulatory system known as negative feedback in which production of a hormone affects a target organ to initiate a response that eventually reduces secretion of that hormone.

Figure 22-3 shows that production of insulin by the β cells of the islets of Langerhans in response to high blood sugar stimulates glucose uptake by skeletal muscle and liver. This process reduces blood sugar and turns off insulin secretion.

PITUITARY

The pituitary, or hypophysis, is called the master gland because it regulates the function of most endocrine organs.

The pituitary is geographically divided into anterior and posterior parts by a line drawn through the middle of the gland. The anterior part consists of the pars distalis and pars tuberalis, and the posterior part consists of the pars nervosa and the pars intermedia. The regions of the pituitary are also classified on the basis of their embryologic origin (see Fig. 22-4).

The adenohypophysis is formed from the oral ectoderm (Rathke's pouch), whereas the neurohypophysis forms as a neuroectodermal downgrowth of the diencephalon.

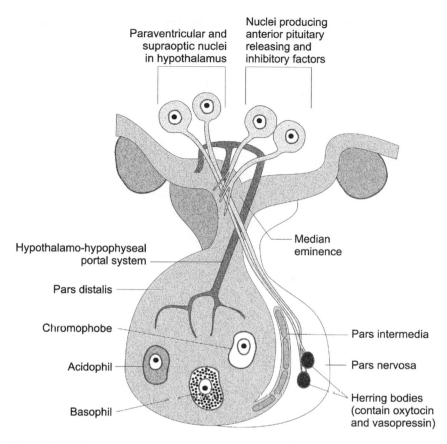

Figure 22-4 Structure of the pituitary.
The macroscopic structure of the pituitary and the relationships of the anterior pituitary to the hypothalamus (hypothalamo-hypophyseal portal system) and the close structural relationship between the hypothalamus and the posterior pituitary (supraoptic and paraventricular neurons terminating as Herring bodies). Not drawn to scale.

The pars distalis communicates with the hypothalamus by the *hypothalamohypophyseal portal system.* A portal system consists of vessels that carry blood from one capillary bed to another. Portal systems may be arterial, as in this case, or venous as in the hepatic portal system.

The portal system is the structural basis for the regulation of pars distalis hormonal secretion by hypothalamic releasing and inhibiting factors.

Releasing and/or inhibiting factors are synthesized in the hypothalamus and released into the hypophyseal portal system, which carries the factors to the sinusoids close to the endocrine cells of the pars distalis. (See table.)

CELLS OF THE PARS DISTALIS*

CELLS	CLASSIFICATION	SECRETION	TARGET OF HORMONE
Somatotrope	Acidophil	Growth hormone (somatotropin, STH)	Stimulates general growth, epiphyses are specific target
Mammotrope	Acidophil	Prolactin (luteotropin, LTH)	Initiates and maintains milk secretion after pregnancy
Thyrotrope	Basophil	Thyroid-stimulating hormone (TSH, or thyrotropin)	Follicular cells of thyroid to produce thyroid hormones
Corticotrope	Basophil	Adrenocorticotropic hormone (ACTH), melanocyte stimulating hormone (MSH)	Adrenal cortex particularly z. fasciculata z. reticularis
Gonadotrope	Basophil	Follice-stimulating hormone (FSH) and luteinizing hormone (LH) from the same cell; LH identical to interstitial cell stimulating hormone (ICSH)	FSH stimulates growth of ova, follices in females and synthesis of androgen-binding protein from Sertoli cells of the testis in men. LH is necessary for ovarian growth, ovulation, and corpus luteum formation in the female and stimulates testosterone production by interstitial (Leydig) cells in males

*Classically, the anterior pituitary cells are classified as *chromophils* (cells which "like" stain) and *chromophobes* (cells which "hate" stain). These classifications represent cells filled with secretory product at the time of fixation (chromophils) and depleted of secretory material (chromophobe).

Cells of the pars intermedia are of little interest in humans, but are believed to be the site of alternative processing of pro-opiomelanocortin (POMC).

> The pars nervosa is unique among glands in that it releases a hormone that is not synthesized in the gland.

The hormones *[oxytocin and vasopressin (antidiuretic hormone)]* are secreted in two nuclei in the hypothalamus (*supraoptic and paraventricular nuclei*). The hormones are bound to binding proteins called neurophysins and transported down the axons of the *hypothalamo-hypophyseal tract* for storage in the pars nervosa. The storage sites are dilated axon terminals known as *Herring bodies*. Oxytocin stimulates the contraction of uterine smooth muscle during the last stages of pregnancy and may be used to induce parturition. Vasopressin has an antidiuretic effect on the kidney. It increases the permeability of the renal distal and collecting tubules causing conservation of water and concentration of the urine. *Diabetes insipidus* is a form of diabetes in which antidiuretic hormone (ADH) is absent and large volumes of water are lost in the urine.

THYROID

The thyroid gland produces two groups of hormones with separate and distinct functions.

Triiodothyronine (T$_3$) and *tetraiodothyronine (T$_4$)* are the *thyroid hormones* and are produced by the principal (follicular) cells (cuboidal and columnar epithelial cells) which surround a central colloid-containing follicle. *Calcitonin*, which is important in Ca^{2+} regulation is produced by the interfollicular (parafollicular, or C) cells (see Fig. 22-5).

The follicular and C cells vary in embryologic origin. The follicular cells are derived from endoderm of the pharynx, whereas the C cells are formed from the ultimobranchial body derived from the fourth pair of pharyngeal (branchial) pouches. *The thyroid gland is the first endocrine gland to develop in the embryo.*

> The precursor of the thyroid hormones is stored in the thyroid follicle.

The colloid consists of *iodinated thyroglobulin*. The thyroglobulin is formed by the follicular cells using the normal secretory pathway and released by exocy-

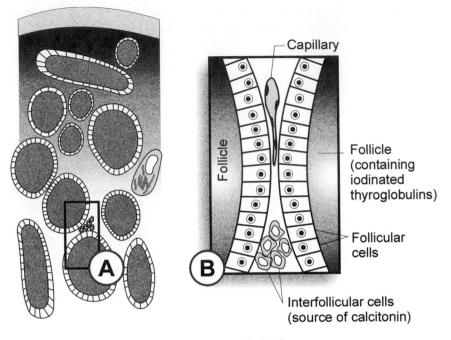

Figure 22-5 Histologic structure of the thyroid gland.
The thyroid is an endocrine gland with two distinctly different hormonal products. The follicular cells and the interfollicular (C or parafollicular) cells both secrete their products into capillaries, but the products are different, T_3 and T_4 in the case of the follicular cells and calcitonin from the C cells. Remember: the colloid in the lumen is not the final secretory product (T_3 and T_4), but is iodinated thyroglobulin which is modified during endocytosis by the follicular cells.

tosis into the follicle. Iodide is taken up the follicular cells and oxidized to form iodine. The iodine is released into the lumen where it iodinates the tyrosine groups of thyroglobulin. Iodinated thyroglobulin is *endocytosed* by the follicular cells and broken down by proteases to form triiodothyronine and tetraiodothyronine (thyroxine), which are released into the bloodstream where they bind to thyroxine-binding protein. C-cell secretion occurs into the bloodstream in response to high Ca^{2+} levels (Fig. 22-5).

As indicated in Chap. 14, *calcitonin* inhibits osteoclastic activity and stimulates osteoblastic activity. It opposes the action of parathyroid hormone (PTH) and reduces blood Ca^{2+} levels.

The thyroid, under control of the pituitary, regulates metabolic rate.

Thyroid hormones are regulated by *thyroid-stimulating (TSH, thyrotropin)* hormone produced by the anterior pituitary, which in turn is controlled by *thyrotropin-releasing factor (TRF)* produced by the hypothalamus. The thyroid hormones regulate metabolism by increasing carbohydrate utilization and influencing such diverse body functions as intestinal absorption, heart rate, and body growth. *Hypothyroidism* results in severe developmental effects in infants *(cretinism) and myxedema* in adults. *Hyperthyroidism* results in *exophthalmic goiter.*

SUPRARENAL (ADRENAL) GLANDS

The adrenals are paired organs located at the cranial pole of the kidneys. The adrenals actually are two separate glands, the adrenal cortex and the adrenal medulla. These regions are structurally distinct, produce different hormones, have separate embryologic origins, and are regulated differently.

The adrenal cortex and medulla are obvious in a gross specimen of adrenal gland. The adrenal cortex produces *mineralocorticoids*, *glucocorticoids*, and

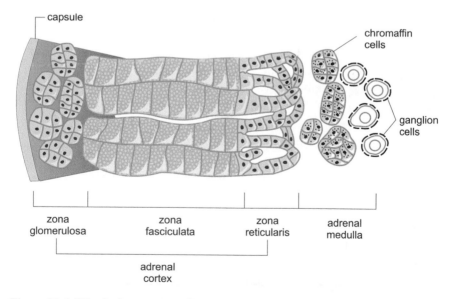

Figure 22-6 Histologic structure of the adrenal gland.
The adrenal gland is two endocrine glands in one organ. The adrenal cortex produces an array of hormones: aldosterone, corticosteroids, and dehydroepiandrosterone. The adrenal medulla functions like a group of modified postganglionic fibers and produces norepinephrine and epinephrine. Ganglion cells are rare.

weak androgens, whereas the adrenal medulla produces *norepinephrine* and *epinephrine*. (See table.) The adrenal cortex forms from the mesodermal epithelium, whereas the adrenal medulla forms from *neural crest*. The adrenal cortex is regulated by the anterior pituitary, and the adrenal medulla is regulated by *preganglionic sympathetic fibers*.

The adrenal cortex uses cholesterol and acetate to form different hormones in each of its zones. Cells of the cortex contain numerous lipid droplets which give them a foamy appearance.

ADRENAL HORMONES

ZONE	SECRETION	TARGET	REGULATORY FACTOR(S)
Zona fasciculata	Mineralocorticoids (mostly aldosterone)	Renal distal and collecting tubules; sweat and salivary gland ducts (Na^+ conserved, K^+ excreted in urine)	Angiotensin II
Zona glomerulosa	Glucocorticoids (mostly cortisol, i.e., hydrocortisone)	Gluconeogenesis by the liver, reduction in cellular protein, mobilization of fatty acids, anti-inflammatory effects	Adrenocorticotropic hormone (ACTH)
Zona reticularis	Weak androgens (dehydroepiandrosterone)	Precursor of estradiol in the fetus, role in adult unclear	ACTH
Adrenal medulla	Norepinephrine and epinephrine	Preparation of body for "fight or flight"	Preganglionic sympathetic fibers

The adrenal gland in the fetus is a key component of the material-fetal adrenoplacental axis providing precursor (dehydroepiandrosterone) for estradiol production in the placenta. The fetal adrenal degenerates after bith leaving the three cortical zones of the adult cortex.

The blood supply of the adrenal is derived from three sources: (1) the adrenal (suprarenal) artery from the aorta, (2) the suprarenal branch of the renal artery, and (3) the suprarenal branch of the inferior phrenic artery, indicating the high degree of vascularity of this gland.

Much of the blood that enters the adrenal passes through the cortex before reaching the medulla. This is critical because the presence of glucocorticoids in the medullary blood supply stimulates the production of epinephrine from norepinephrine through a methylation mechanism.

Species with separate adrenal cortex and adrenal medulla produce a higher concentration of norepinephrine compared with the predominance of epinephrine in mammals where the adrenal cortical secretion of glucocorticoids stimulates methylation by a methyl transferase enzyme.

PARATHYROID GLANDS

Serum Ca^{2+} levels are primarily regulated by PTH, which responds to low serum Ca^{2+} by stimulating osteoclastic and bone-lining cell activity.

The principal cell of the parathyroid produces PTH, which responds to low Ca^{2+} levels by stimulating *osteoclasts to resorb bone* and *bone-lining cells to transport Ca^{2+}* from the bone fluid to the extracellular fluid. The result is an increase in blood Ca^{2+} levels. PTH is the primary regulator of Ca^{2+} levels in the blood. Although calcitonin from the parafollicular (C) cells of the thyroid opposes the action of PTH, it is not an equal opponent (see Chap. 14). *Oxyphil cells* are found in groups around the parathyroid gland, increase with age, and appear to be derived from the principal cells. Their function is unknown, although oxyphilomas produce PTH.

The parathyroid glands are derived embryologically from the third and fourth branchial pouches and are four separate glands located on the posterior capsule of the thyroid gland. These glands and their secretion of parathyroid hormone (PTH) are independent of pituitary regulation. Calcitonin secretion from the thyroid is also independent of pituitary regulation.

PINEAL GLAND

The pineal gland contains two main cell types: pinealocytes which secrete melatonin, a hormone involved in maintenance of circadian rhythms, and glial cells, which are supportive. The pineal gland is also involved in regulation of reproductive function as part of a hypothalamo-pituitary-gonado-pineal axis.

Melatonin from the pineal gland is released into the bloodstream. Secretion is regulated by postganglionic sympathetic fibers from the superior cervical ganglion. The pineal gland contains *brain sand* (corpora arenacea, or acervuli, concretions of calcium carbonate and other phosphates), which increases with age and is a landmark in radiologic examinations of the region because of the calcification.

The pineal gland (epiphysis) is closely associated with the CNS (attached by a stalk to the roof of the third ventricle) and arises as a downgrowth of the diencephalon during fetal development.

OTHER GLANDS

All the glands discussed to this point are merocrine glands; no membrane is lost as the cells release secretory product by exocytosis. In other glands, part of the cell (apocrine) or the whole cell (holocrine) is lost as the secretion is released.

The sebaceous glands discussed in Chap. 23 are holocrine glands. Most sweat glands are classified as merocrine. However, those found in the axilla, circumanal, labia majora, and areola (nipple of the breast) are apocrine glands in which the cells release part of their apical cytoplasm during secretion. These glands are influenced by reproductive hormones and become functional only after puberty. The mammary glands (Chap. 25) are highly modified sweat glands which function as exocrine glands utilizing the lactiferous duct system to release milk in response to suckling during lactation. The secretory cells of the mammary gland release milk proteins by exocytosis, but lipids are released at the apex of the cell with the loss of apical cytoplasm (Chap. 25).

INTEGUMENTARY SYSTEM

·

- **Development of the Skin**
- **Keratinocytes**
- **Pigmentation of the Skin**
- **Nonkeratinocytes in the Epidermis: Melanocytes**
- **Langerhans Cells**
- **Merkel Cells**
- **Dermis**
- **Hair**
- **Sebaceous Glands**
- **Sweat Glands**
- **Nails**

·   ·   ·   ·   ·   ·   ·   ·   ·   ·   ·   ·

The skin—it's what keeps everything in.

The integument, or integumentary system, encompasses the skin as well as specialized structures (appendages) derived from the skin. The hair, nails, and even teeth are often included as parts of the integumentary system. The teeth are discussed in Chap. 21, "Gastrointestinal System."

The skin is important for protection of the body from the external environment. For this reason (teleologic), it forms early during development.

DEVELOPMENT OF THE SKIN

The skin forms from ectoderm and the underlying mesenchyme. The epidermis is a stratified squamous epithelium that is derived from ectoderm. The dermis is formed from the underlying mesenchyme and contains blood vessels and the usual connective tissue components.

During development the skin first appears as a single layer of ectodermal cells. These cells divide to form an additional outer layer of flattened cells called the *periderm* that combines with amniotic fluid and the biologic garbage floating therein (hair, urine, and sebum) to form a protective coating for the embryo called the *vernix caseosa*. During later development the basal layer continues to proliferate and eventually differentiates into the layers observed in adult skin: *stratum germinativum, stratum spinosum, stratum granulosum, stratum lucidum,* and *stratum corneum* (Fig. 23-1).

KERATINOCYTES

The epidermis is composed primarily of cells called keratinocytes. These cells originate in the stratum (s.) germinativum (basale) and move upward toward the surface as they differentiate.

Intermediate filaments (*cytokeratins*) are observed in the cytoplasm and increase in volume as the keratinocyte differentiates and moves upward toward the surface (Fig. 23-1). Mitotic figures are observed in the s. basale; therefore this is the layer affected by chemotherapeutic or radiation treatments. In a single

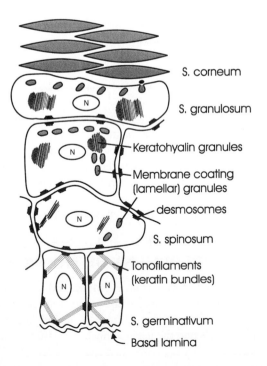

S. corneum

S. granulosum

Keratohyalin granules

Membrane coating (lamellar) granules

desmosomes

S. spinosum

Tonofilaments (keratin bundles)

S. germinativum

Basal lamina

Figure 23-1 The histologic layers of the epidermis.
Cells born in the stratum germinativum (basale) ascend through the epidermal layers and are sloughed off at the surface. The keratohyalin and membrane-coating granules develop in the stratum spinosum. Tonofibrils become highly developed in the stratum spinosum.

layer, these basal cells are bound to each other by *desmosomes* and to the basal lamina by *hemidesmosomes*. The s. basale and the next layer, the s. spinosum (spiny layer), are sometimes lumped together as the s. Malpighii. The s. spinosum consists of several layers of cells with abundant *desmosomes* and intercellular connections formed by *tonofibrils*. These interconnecting bridges resist abrasion and maintain cohesiveness between the keratinocytes. The next layer is the granulosum (granular layer), which consists of flattened cells that accumulate *keratohyalin granules* and *membrane-coating (lamellar) granules* (Fig. 23-1).

The keratohyalin granules provide the matrix, particularly the protein *filaggrin*, that facilitates the embedment and aggregation of keratin filaments.

Filaggrin is an intracellular protein which is involved in the arrangement (compaction) and cross-linking of the keratin.

The epidermis is impermeable to water in both directions. The impermeability of the epidermis is established by the contents of the lamellar granules at the level of the granular layer.

The lamellar granules are packaged on leaving the Golgi apparatus and fuse with the plasma membrane before release of their contents into the matrix surrounding the keratinocytes in the s. granulosum.

The *stratum lucidum* is detected only in thick skin. It contains *eleiden*, which is believed to be responsible for the refractility of this layer. The s. corneum is the outermost layer of the epidermis and like the s. spinosum is thicker in thick skin. The layers of the s.corneum are composed of dead (nonnucleated) squames or scalelike cells that become progressively flattened. The cytoplasm of the cells is replaced by keratin, the formation of which is initiated in the more basal layers.

PIGMENTATION OF THE SKIN

Skin color is dependent on carotene, blood, and melanin. Skin color differences are due to differences in the number of melanin granules.

Pigmentation of the skin is based on a number of different factors. The skin color is due to a combination of yellow from *carotene*, a pigment deposited in the s. corneum; red from *blood* within the dermal vessels; and brown from the *melanin* present in the spiny and germinative layers. Melanin is found within melanocytes that establish a relationship with neighboring keratinocytes. In specific regions of the body there is a constant ratio of melanocytes per unit area. Differences in skin color are *not* related to the number of melanocytes, but are due to the quantity of melanin granules found within keratinocytes.

NONKERATINOCYTES IN THE EPIDERMIS: MELANOCYTES

Melanocytes are derived from neural crest cells and utilize tyrosinase to convert tyrosine into dopa, a precursor of melanin.

Pre-melanocytes (neural crest cells) migrate into the epidermis during development. Melanin is formed within *tyrosinase*-containing melanosomes by the conversion of tyrosine into 3,4-dihydroxyphenylalanine (*dopa*), then dopaquinone, and subsequently melanin. The melanin granules pass along the cytoplasmic extensions of the melanocyte and are injected into keratinocytes. This relationship between melanocytes and keratinocytes is known as the *epidermal-melanin unit*. Melanocytes and other nonkeratinocytes are shown in Fig. 23-2.

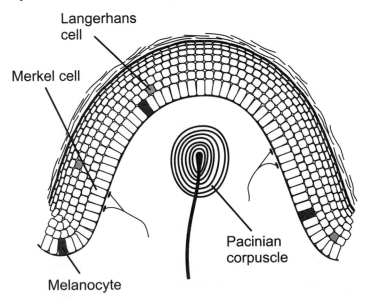

Figure 23-2 A higher magnification view of integumentary histology.
The diagram indicates the position of the nonkeratinocyte cells of the epidermis. The melanocytes, Merkel cells, and Langerhans cells are shown in the epidermis with an underlying Pacinian corpuscle (in the dermis) for detection of deep pressure.

LANGERHANS CELLS

Langerhans cells are the antigen-presenting cells of the skin.

The Langerhans cells are derived from the monocyte bone marrow lineage. They function as the macrophages of the skin and are capable of phagocytosis and antigen presentation. Langerhans cells secrete and respond to small peptide or protein intercellular messengers called cytokines. The Langerhans cells contain strange tennis racquet shaped cytoplasmic inclusions (Birbeck granules) and are not attached to neighboring keratinocytes by desmosomes.

MERKEL CELLS

Merkel cells are mechanoreceptors in the skin.

These cells are usually found in the s. germinativum close to intracpithelial nerve endings (Fig. 23-2). Merkel cells are linked by desmosomes to their neighboring keratinocytes. They contain catecholamine-like granules, and the cells are believed to function like *mechanoreceptors*.

DERMIS

The dermis forms from mesenchyme, induces development of epidermis and its appendages, and provides support for the epidermis.

The dermis forms embryologically from the dorsal-most part of the somite (*dermatome*) and some somatic mesoderm. The mesenchyme of the dermis drives the differentiation of the overlying epidermis. Histologists often divide the dermis into a papillary layer and a deeper reticular layer. The papillary layer provides anchoring for the epidermis, and the reticular layer contains more collagen and *elastin* and provides the mechanical strength of the skin. The dermis is filled with a glycosaminoglycan matrix, which creates a gel-like environment within the dermis.

Glycosaminoglycan content and therefore the gel-like state of the dermis increase with age. Fibroblasts, which synthesize the collagen and elastin fibers, are the predominant cell type along with numerous macrophages. Smooth muscle fibers are found in the dermis associated with hair follicles as the arrector pili muscles. Underlying the reticular layer is a layer of subcutaneous tissue known as the *hypodermis*.

HAIR

Hair develops from the epidermis as elastic, keratinized threads. Each hair consists of a root and free shaft surrounded by the hair follicle.

The hair follicle consists of an epithelial (epidermal) part and a connective tissue (dermal) part.

Sebaceous glands and arrector pili are associated with the hair follicles. The lower end of the hair follicle forms an expanded bulb-like ending indented by the connective tissue papilla. This arrangement allows for continuous epithelial-mesenchymal interactions. The hair itself consist of a central pigment-containing medulla surrounded by a multilayer cortex and a single-layered cuticle. These layers differ in pigmentation and arrangement in different races and different hair colors.

The hair follicle is derived partially from the dermis (the external dermal root sheath) and partially from epidermis (the internal epithelial root sheath).

The internal root sheath has three layers from outside to inside: Henle's layer (flattened clear cells), Huxley's layer (trichohyalin granules analogous to keratohyalin granules and tonofibrils), and the cuticle of the root sheath (transparent scale-like). The hair grows through mitosis of epidermal cells induced by the dermal papilla. All hairs go through three stages on the way to hair loss: a growth phase (anagen), a termination phase (catagen), and the resting phase (telogen) which precedes hair loss.

SEBACEOUS GLANDS

Sebaceous glands continuously produce sebum, which is released into the hair follicle.

Sebaceous glands are holocrine glands (the entire cell is lost on secretion, Chap. 22) which release an oily material, called sebum, continuously into a duct that opens into the hair follicle. Growth of the sebaceous glands is stimulated at puberty by the sex hormones in both males (androgen) and females (estrogen).

SWEAT GLANDS

There are both merocrine and apocrine sweat glands.

Sweat glands consist of coiled secretory portions lined by a simple epithelium, whereas the duct system is a stratified cuboidal epithelium. Most sweat glands are the merocrine type, although those in the labia majora, areola, and axillary and anal regions are classified as apocrine in which the apex of the cell is lost. The secretion from these glands is thicker and more viscous than the merocrine-type secretion from "normal" sweat glands. The innervation of the two types of sweat gland differs: adrenergic endings innervate the apocrine glands, whereas cholinergic endings innervate the merocrine glands.

NAILS

The nails develop from the epidermis, which covers the dorsal surface of the terminal phalanx and invades the underlying mesenchyme. The epithelium forms a nail plate, which is cornified. Nail growth occurs from the nail bed, which feeds cells into the growing nail. The other parts are the *hyponychium*, which is the thickened stratum corneum of the epidermis at the junction of the epidermis of the nail bed and the epidermis of the fingertip, and the *eponychium*, or cuticle.

URINARY SYSTEM

·

- **The Renal Corpuscle**
- **The Filtration Apparatus**
- **Modification of Provisional Urine**
- **Transitional Epithelium**
- **Clinical Correlates**

· · · · · · · · · · · ·

The key word for understanding the kidney is *homeostasis*, particularly of fluid and electrolyte levels. The human kidney consists of millions of mini-organs, the nephrons, which are attached to a confluent drainage system of collecting tubules and ducts. The individual parts of the nephron and collecting system have specific roles in the manufacture of hypertonic (concentrated) urine. Because it exports a product (urine) through a duct system to the external environment, the kidney could be considered an exocrine gland. However, the kidney also plays a critical role as an endocrine organ in the regulation of blood pressure.

THE RENAL CORPUSCLE

The first part of the nephron, the renal corpuscle (Bowman's capsule) is involved in making a filtrate of blood plasma. The remainder of the nephron plus the collecting system is involved in modifying that filtrate to produce a hypertonic urine.

Bowman's capsule (Fig. 24-1) is like many of the other epithelium-lined body cavities, only on a reduced scale. For instance, the abdominal cavity is an empty sac lined by a simple squamous epithelium (the peritoneum). The organs of the abdomen are not within the peritoneal cavity, but rather push into its space covered by a layer of peritoneum. In similar fashion, a tuft of capillaries, the *glomerulus*, indents one side of the renal capsule and becomes intimately associ-

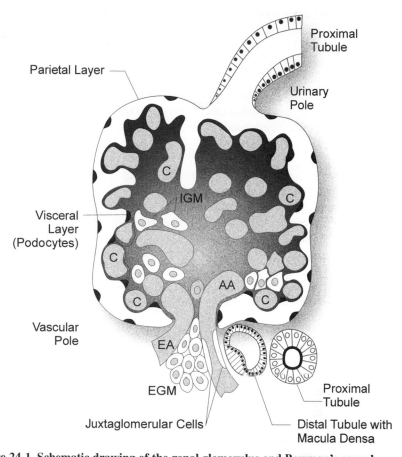

Figure 24-1 Schematic drawing of the renal glomerulus and Bowman's capsule.
The parietal layer of the capsule consists of a simple squamous epithelium which is continuous at the vascular pole with the single layer of specialized cells (podocytes) forming the visceral layer of the capsule. The glomerulus consists of a convoluted tuft of capillaries (C) and intraglomerular mesangial cells (IGM) (some capillaries and IGM have been omitted from the drawing for clarity). Extraglomerular mesangial cells (EGM) are found at the vascular pole between the efferent (EA) and afferent (AA) arterioles.

ated with some of the capsule's lining cells. This is the *visceral layer* of Bowman's capsule. It is continuous with a layer of cells (*parietal layer*) defining the outside of the capsule. Between the two layers is the *urinary space*. Consider a large, soft beanbag chair. A person who sits in it is completely engulfed by it, leaving only arms and legs protruding. The part of the chair in contact with the body is the visceral layer. It becomes the outside surface of the chair at the place where the arms and legs stick out. The body represents the tuft of capillaries, and the arms and legs represent the blood vessels entering (*afferent arterioles*) and leaving (*efferent arterioles*) the glomerulus. The space between the inner and outer layers of the chair is the urinary space, which in the glomerulus drains into the proximal tubule of the nephron at the urinary pole of the glomerulus.

> The glomerular filtration apparatus of the nephron consists of the capillary endothelium, the specialized cells of the visceral layer of the capsule, and a shared basement membrane between them.

The visceral layer of Bowman's capsule (Fig. 24-1) consists of a single layer of specialized cells (*podocytes*) with numerous *primary processes* which surround the capillary walls. These processes give off secondary processes (*pedicels*) which interdigitate with pedicels from adjacent podocytes, forming *filtration slits* between them which are bridged by a thin membrane of protein (Fig. 24-2). In cross section, the effect is a line of pedestals linked by a thin rope. The endothelium of the capillary is *fenestrated*, allowing the passage of fluid, ions, and some proteins, but not cells. Between the capillary endothelium and the pedicels of the podocytes is a shared *glomerular basement membrane (GBM).*

Figure 24-2 The glomerular filtration apparatus.
The filtration apparatus consists of a fenestrated capillary endothelium, a basement membrane, and the slits between adjacent processes (pedicels) of podocytes. The glomerular basement membrane consists of inner and outer lamina rarae containing high concentrations of negatively charged glycosaminoglycans and a central lamina densa containing type IV collagen. The surface of the pedicels contains podocalyxin, a polyanion rich in sialic acid residues.

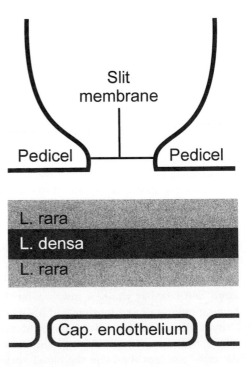

THE FILTRATION APPARATUS

The glomerular basement membrane is the main filtration barrier and consists of an electron-dense central core separating two electron-lucent layers.

The central *lamina densa* consists primarily of *type IV collagen* and laminin and presents a structural barrier to the passage of macromolecules (Fig. 24-2). The *lamina rara externa* (adjacent to the pedicels) and *lamina rara interna* (adjacent to the endothelial cells) contain fibronectin and a mixture of negatively charged (polyanionic) *glycosaminoglycans* including *heparan sulfate* which bind cations and repel anions. The GBM is assisted by *podocalyxin*, a sialic acid-rich polyanion on the surface of the pedicels. The filtration apparatus allows the passage of proteins smaller than albumin (70 kD). Thus, urinary albumin content is useful in determining the integrity of the GBM.

The driving force for filtration is the blood pressure in the glomerular capillaries (a high-pressure capillary bed). This filtration force has two opposing pressures, the colloid osmotic pressure of the blood (produced primarily by albumin) and the hydrostatic pressure of the urinary space, resulting in a net filtration force of about 10 to 15 mmHg (torr). Thus, the rate of glomerular filtration can be affected by blood pressure. One way in which atrial natriuretic peptide (ANP, see Chap. 18) increases water output by the kidneys is through a differential effect on intraglomerular pressure. As in most vascular beds, ANP dilates the afferent arteriole allowing more blood into the glomerulus. However, ANP also constricts the efferent arteriole, allowing less blood to leave the glomerulus, resulting in increased perfusion pressure and a larger volume of filtrate.

Mesangial cells are found within the glomerulus and at the vascular pole.

Intraglomerular mesangial cells (IGM) (Fig. 24-1) span the distance between adjoining capillaries in the glomerulus and cover those portions of the capillary endothelium not covered by podocytes. These cells provide structural support to the capillary tuft, secrete some components of the extracellular matrix, and maintain the GBM by phagocytosis. In addition, these cells share some similarities with pericytes (e.g., actin and myosin bundles) and may play a role in regulation of glomerular blood flow. A second group of mesangial cells, the *extraglomerular mesangium (EGM)*, is located between the afferent and efferent arterioles at the vascular pole.

MODIFICATION OF PROVISIONAL URINE

The remainder of the nephron (proximal tubule, thin loop, distal tubule) modifies the provisional urine produced in Bowman's capsule (Fig. 24-3).

The Proximal Tubule

Reabsorption of most of the components of urine (including water) occurs in the proximal tubule.

Unmodified (provisional) urine enters the proximal tubule of the nephron from the urinary space at the urinary pole of the capsule (Fig. 24-1). Proximal tubule cells are the reverse filters of the nephron. They remove virtually all the glucose, amino acids, proteins and complex carbohydrates from the urine along

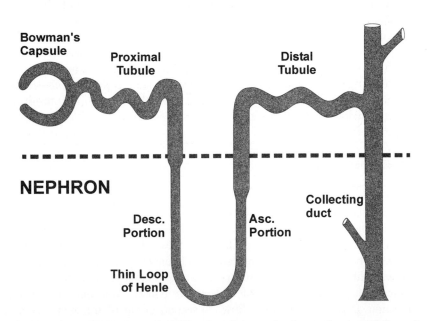

Figure 24-3 Simplified schematic of Bowman's capsule, the nephron, and the collecting duct.
Functions of the specific regions are described in the text.

with 85 percent of the Na^+, Cl^-, and H_2O using a complex system of transporters, co-transporters, channels, and exchanger proteins.

> The driving force for reabsorption of the filtrate is the presence of sodium pumps (Na^+, K^+-ATPase) in the basolateral cell membranes.

The pumping capacity of the proximal tubule cells is greatly aided by increased membrane surface area provided by the numerous basolateral infoldings and interdigitations between adjacent cells. The Na^+, K^+-ATPase pumps sodium out of the cell (in exchange for K^+) into the intercellular space. The osmotic gradient formed by the high concentration of sodium in the intercellular space draws water (containing dissolved Cl^- and amino acids) from the lumen. Glucose is most likely transported to the intercellular space through a glucose-Na^+ exchanger. The resulting hydrostatic pressure forces the water and dissolved substances into the peritubular capillary network.

> Proteins (<70 kD) and complex carbohydrates are broken down by surface enzymes, and the products are endocytosed by the proximal tubule cells for further digestion.

The lumenal surface area of proximal tubule cells is greatly increased by numerous *microvilli*. The *glycocalyx* associated with the microvilli contains many enzymes which partially digest proteins and carbohydrates. The breakdown products enter an extensive apical canalicular system from which endocytic vesicles bud to transport these fragments to the lysosomes for further digestion. (See Chap. 6, "Intracellular Trafficking").

> Proteins and other substances too large to pass the GBM may enter the urine by a different mechanism.

Proximal tubule cells remove some metabolites and foreign substances from the blood and secrete them into the urine. Substances secreted into the urine by the back door include para-amino hippuric acid (PAH) and penicillin. The *clearance rate* for radiolabeled substances defines the speed at which these substances are removed. In the case of the proximal tubule, this represents material transported from the blood across the proximal tubule cells into the urine. Clearance rates can be used to judge the "health" of the proximal tubule. Clearance rates can also be calculated for glomerular filtration using endogenous (creatinine) or exogenous (inulin) substances which are just filtered and not further processed by the tubule segments.

The Thin Loops

> The thin loops (of Henle) regulate the volume and tonicity of urine in the collecting ducts.

Nephrons located near the periphery of the renal cortex generally have short thin loops which may never extend into the inner medulla, whereas those near the medulla (*juxtamedullary*) have thin loops which reach deep into the tip (papillary portion) of the medulla. The *descending limb of the thin loop* (Fig. 24-3) is permeable to H_2O, Na^+, and Cl^-. However, the *ascending limb of the thin loop* is *impermeable to H_2O* and actively pumps Cl^- out of the urine into the interstitial space of the medulla. Na^+ follows passively to maintain the ionic gradient. The impermeability of the ascending thin limb may be due to the absence of *water channel-forming proteins* such as *aquaporin*, which is expressed in the proximal tubule and thin descending limb, but not in the ascending limb, Na^+ and Cl^- in the interstitial space may reenter the descending limb of the thin loop but are kicked out again in the ascending limb (*countercurrent exchanger*). This has two effects:

1. Delivery of a hypotonic filtrate to the distal tubule
2. Establishment of an increasing concentration gradient of Na^+ which is highest at the tip of the thin loop (deep in the medulla). Thus, as the collecting ducts pass through the medulla toward the tip of the papilla, they encounter an ever-increasing osmotic gradient which draws more and more water out of the duct, resulting in a hypertonic urine. The excess water is removed by capillary loops, the vasa recta, which extend deep into the papilla. The permeability of the collecting ducts to water is regulated by *antidiuretic hormone (ADH)*.

The Distal Tubule

> Aldosterone regulates Na^+ resorption from urine in the distal tubule.

The cells of the distal tubule (Fig. 24-3) have extensive basal infoldings which increase the membrane surface area and enclose numerous mitochondria which provide ATP for local Na^+, K^+-ATPase pumps. These cells are the Arnold Schwarzeneggers of ion pumping, removing sodium from the urine and thus making it more hypotonic. This process is regulated by *aldosterone*, a mineralo-corticoid secreted by the zona glomerulosa of the adrenal cortex. Aldosterone secretion is regulated by the status of body fluid volume (expanded or depleted) via *atrial natriuretic peptide* (inhibits aldosterone secretion) and *angiotensin II* (stimulates aldosterone secretion, see below).

> The macula densa of the distal tubule plays a special role in fluid-elec-trolyte homeostasis and blood pressure regulation.

Where the distal tubule passes close to the vascular pole of its glomerulus, its morphology is altered. A segment of its normally cuboidal cells become elon-gate and densely packed to form the *macula densa* (Fig. 24-4). In addition, the polarity of these cells is reversed with the nucleus lying near the lumenal end and the new apex of the cell aimed at the specialized smooth muscle (*juxtaglomeru-lar*) cells of the afferent arteriole. Current evidence suggests that cells of the

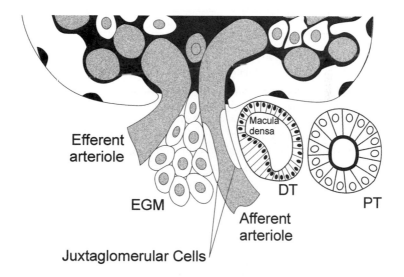

Figure 24-4 The vascular pole of the glomerulus.
Juxtaglomerular cells (JG) are specialized smooth muscle cells of the afferent arteriole. JG cells secrete the enzyme renin, which converts angiotensinogen into angiotensin I, the pre-cursor to angiotensin II, a potent vasoconstrictor. The activity of JG cells is regulated by cells of the macula densa located in a portion of the distal tubule (DT), which passes close to the vascular pole. EGM, extraglomerular mesangial cells; PT, proximal tubule.

macula densa monitor the urine (perhaps Cl⁻ content) and signal the juxta-glomerular (JG) cells (possibly by NO release). This process is known as *tubu-loglomerular feedback*. When stimulated, the JG cells secrete *renin* into the blood. Renin is an enzyme which converts *angiotensinogen* (secreted by the liver) into the decapeptide, *angiotensin I*. In turn, angiotensin I is converted to the octapeptide angiotensin II by *angiotensin converting enzyme (ACE)* present on the surface of endothelial cells in most arterial vascular beds, but at highest con-centration in the pulmonary vasculature. As noted above, angiotensin II stimu-lates the secretion of aldosterone, which increases Na^+ resorption (and thus water retention) from the distal tubule. Angiotensin II is also a potent vasoconstrictor, which reduces blood flow to the renal glomeruli resulting in decreased glomeru-lar filtration and thus decreased volume of the provisional urine entering the proximal tubule. (See opposite effects of ANP, page 333.) Angiotensin II increases blood pressure by all these mechanisms.

The Collecting Ducts

> Final adjustments to urine volume are made in the collecting ducts under the regulation of antidiuretic hormone from the posterior pituitary.

> Transmembrane proteins (aquaporins) in the membrane of collecting duct epithelial cells form water channels.

At least two types of *water channels* are found in the plasma membrane of collecting duct cells. Those located on the basolateral membrane, aquaporins 3 and 4, are constitutive (i.e., always present). In the apical (lumenal) membrane, however, it appears that another aquaporin protein (aquaporin 2) aggregates to form open channels only under the influence of ADH. When these channels are open, water can flow from the lumen, across the collecting duct cells, into the interstitium under the influence of the osmotic gradient produced by the thin loop (see above).

TRANSITIONAL EPITHELIUM

Transitional epithelium (urothelium) contains a vesicle shuttle system which maintains the water impermeability of the surface cells.

Transitional epithelium is found only in the urinary system. It lines everything from the *minor calyces* to the prostatic urethra, including the bladder. It is a specialized epithelium (see Chap. 12) consisting of several layers of cells which respond to stretch produced by liquid filling the passageway, particularly in the urinary bladder. The surface cells of the *urothelium* in the undistended bladder contain abundant vesicles in the apical cytoplasm. The flattened vesicles contain a high content of the lipid cerebroside, have very particulate sections connected by short portions of thin membrane, and look similar to a closed clam shell. When the bladder is distended, these vesicles move to the surface and fuse with the apical membrane, adding areas of paracrystalline protein arrays to the outer surface.

CLINICAL CORRELATES

The effects of diabetes mellitus and hypertension result in significant renal pathology. Diabetes results in thickening of the GBM (glycation) and leakage of albumin into the urine (hyperalbuminuria). Diabetes mellitus is associated with glomerula hypertension and hyperfiltration. Hyperinsulinemia resulting from non-insulin dependent diabetes increases renal sodium retention and activity of sympathetic neurons leading to systemic hypertension.

REPRODUCTIVE SYSTEMS

·

- **Female Reproductive System (E Pluribus, Unum)**
- **Uterine/Ovarian Regulation**
- **The Vagina and Cervix**
- **Mammary Glands**
- **Male Reproductive System (Out of One, Many)**
- **The Testis**
- **Male Reproductive System: Ducts and Accessory Glands**
- **Fertilization**
- **Placenta**

·   ·   ·   ·   ·   ·   ·   ·   ·   ·   ·   ·

Sexual reproduction is advantageous for survival in an environment where conditions may change without warning. Sexual reproduction is beneficial in that it allows both reorganization of maternal and paternal genomes as well as recombination of the genomes through phenotypic selection. The male and female reproductive systems are designed to facilitate genetic recombination in the development of germ cells (meiosis) and the union of male and female gametes to produce a zygote (fertilization). Other components of the female and male reproductive systems play a supportive role in these critical events.

FEMALE REPRODUCTIVE SYSTEM (E PLURIBUS, UNUM)

The role of the female reproductive system is twofold: to produce a single haploid gamete on a regular basis between puberty and menopause and to prepare to receive and maintain a conceptus (also on a regular basis). The

first role (ovarian cycle) concerns the development, maturation and, ultimately, expulsion of a single oocyte from the ovary (Fig. 25-1). The second role involves the cyclical priming of the uterine bed to receive a fertilized oocyte (conceptus or embryo) and, if fertilization does not occur,
returning the uterine bed to the starting point (menstrual cycle).

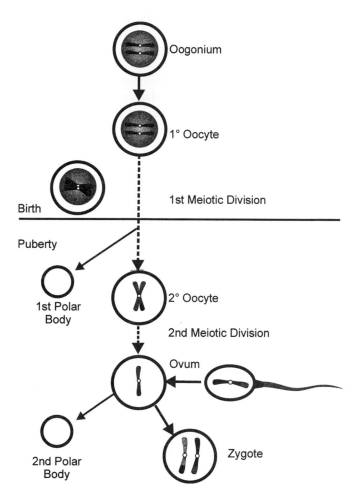

Figure 25-1 Oogenesis.
Oogonia proliferate and enter the first meiotic division before birth. They become arrested
in prophase of the first meiotic division until they resume individually at puberty.
Exchange of material between homologous chromosomes occurs in the first meiotic division. For a detailed review of meiosis, see Figs. 10-12 and 10-14.

UTERINE/OVARIAN REGULATION

Both the ovarian and uterine cycles are under control of hormones released
from the anterior pituitary and ovary.

Under the cyclical regulation of releasing hormones (GnRH) from the hypothalamus (see Chap. 22), gonadotrophs of the anterior pituitary release *follicle
stimulating hormone (FSH)* and *luteinizing hormone (LH)*. *Leptin*, the adipocyte
hormone, also regulates both GnRH and FSH/LH release, probably as a mechanism to correlate reproductive ability with nutritional status. FSH primarily regulates events leading to ovulation such as follicular development and the production of ovarian estrogen, whereas LH stimulates ovulation and the production of
ovarian progesterone. A dramatic increase in LH secretion (the *LH surge*) occurs
24 h in advance of ovulation.

Primordial germ cells give rise to oocytes. Usually, only one oocyte completes development during each ovarian cycle.

Embryonic primordial germ cells originate in the yolk sac and migrate into
the gonadal ridge. Within the developing gonad, primordial germ cells proliferate as diploid germ cells (oogonia). Before birth, a large number of oogonia enter
prophase of the *first meiotic division* but become arrested in the dictyate stage
until ovulation.

Follicular development is regulated by FSH. The transition from primordial follicle to mature secondary (Graffian) follicle takes several cycles and
is a continuous process.

The oocytes are each surrounded by a single layer of flattened follicular cells
which develop into a supportive layer for the oocyte, providing nutrients and
removing waste. Together, they are referred to as primordial follicles. As the primordial follicles awaken, the oocytes grow in size and the follicular cells become
cuboidal and proliferate to form a multilayered primary follicle surrounding a primary oocyte. The follicular cells are separated from the surrounding ovarian connective tissue (*stroma*) by a basement membrane, whereas the oocyte is separated
from the follicular cells by the *zona pellucida*, a glycoprotein coat produced
mainly by the oocyte. The follicular cells and oocyte remain in intimate contact

via gap junctions located on cellular processes which pierce the zona pellucida. The gap junctions allow the passage of only small molecules which may serve as precursors to macromolecules synthesized in the oocyte. During subsequent development, the follicular cells differentiate into *granulosa cells*, which secrete *estrogens*. Follicular cells may also secrete substances (e.g., activins and other factors) which assure that only a single oocyte is normally ovulated from the pair of ovaries.

As the oocyte makes the transition into a secondary follicle, small spaces develop between the granulosa cells and coalesce to form a fluid-filled space (antrum) within the follicle. Enlargement of the antrum presses the granulosa cells to the periphery of the follicle and leaves the oocyte with a few surrounding cells (*corona radiata*) resting on a pillow of granulosa cells, the cumulus oophorus (Graffian follicle) (see Fig. 25-2). Stromal cells surrounding the follicle differentiate into cells of the *theca interna* and *theca externa*, the former acquiring the characteristics of steroid-producing cells and the latter retaining their fibroblast-like nature. Theca interna cells secrete *androstenedione* (see below). When ovulation occurs, the oocyte completes the first meiotic division, gives off the

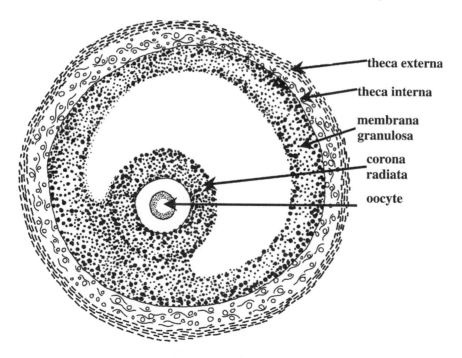

Figure 25-2 Late secondary (Graffian) follicle.
A secondary oocyte is surrounded by cells of the corona radiata. The walls of the follicle consist of granulosa cells and two layers of thecal cells. After ovulation, granulosa (lutein) cells will produce progesterone.

first polar body, and proceeds to the second meiotic division where it becomes arrested until fertilization occurs (Fig. 25-1).

The raison d'etre of granulosa cells is to produce estrogens.

Granulosa cells possess all of the enzymes necessary for the production of *estradiol* (estradiol-17β). They have the typical appearance of other steroid-producing cells including abundant smooth endoplasmic reticulum (SER), intracellular lipid droplets, and mitochondria with tubular cristae. Granulosa cells also have the capability of converting the androstenedione produced by the theca interna cells to estradiol.

Estrogen synthesis increases during follicular maturation and plays numerous crucial roles in ovulation and the menstrual cycle.

Below is a partial list of the roles of estrogens:

1. Estrogens prime the thecal cells to become lutein cells and produce progesterone.
2. Estrogens stimulate the LH surge from the anterior pituitary and decrease FSH production.
3. Estrogens stimulate synthesis and secretion of lytic enzymes, prostaglandins, plasminogen activator (tPA), and collagenase, which participate in the rupture of the ovarian wall and follicle immediately prior to ovulation (release of a single oocyte).
4. Estrogens promote increased height and ciliation of oviduct epithelial cells to facilitate transport of the ovum.
5. Estrogens stimulate proliferation of epithelial cells in the glands of the uterine wall leading to repopulation of the uterine lining (endometrium) after menses.
6. Estrogens stimulate synthesis and storage of glycogen in the endometrium and vaginal epithelium.

Estrogens participate in *negative feedback* regulation of their own synthesis.

Estrogens inhibit secretion of FSH from the anterior pituitary by two mechanisms: (1) a direct inhibitory effect on gonadotrophs and (2) inhibition of GnRH secretion from the hypothalamus. Decreased release of FSH leads to reduced stimulation of follicular development and therefore reduced synthesis and secretion of estrogens. Negative feedback is discussed in Chap. 22.

> The LH surge induced by rising blood estrogen levels induces ovulation (expulsion of the oocyte), resumption of the first meiotic division to form the secondary oocyte, and transformation of the post-ovulational follicle into the corpus luteum.

Both the granulosa (g. lutein) cells and theca interna (t. lutein) cells of the ovulated follicle involute to form the *corpus luteum*. The theca lutein cells form a thin cortex surrounding a core of granulosa lutein cells. A blood clot resulting from ovulation may persist at the center of the corpus luteum until the clot is replaced by connective tissue. The cells of the corpus luteum produce *progesterone* which, as its name suggests, prepares and maintains the endometrium for implantation of the embryo and subsequent gestation.

> Progesterone initiates and maintains the endometrium in the secretory phase of the menstrual cycle. If implantation fails to occur, menses is initiated as LH levels decline.

Menstruation (days 1 to 4 of the menstrual cycle) results in the loss of most of the surface epithelium (endometrium, *stratum functionalis*) from the uterine wall. Only the epithelium of the basal portion of the uterine glands [*stratum* (s.) basalis], remains (Fig. 25-3). During the *proliferative phase* (days 5 to 14), estrogen produced by the ovarian follicle stimulates the proliferation of gland cells in s. basalis, resulting in repopulation of the surface endometrium and regrowth of the straight glands. At this time, coiled arteries begin to penetrate the new stratum (*s.*) functionalis.

The *secretory phase* (days 15 to 28) is stimulated by progesterone produced by the corpus luteum. The secretory glands of the endometrium become tortuous, accumulate glycogen, and secrete glycoproteins which will nourish the embryo should fertilization and implantation occur. After implantation, nutritional requirements of the embryo are supplied by the mother through the placenta.

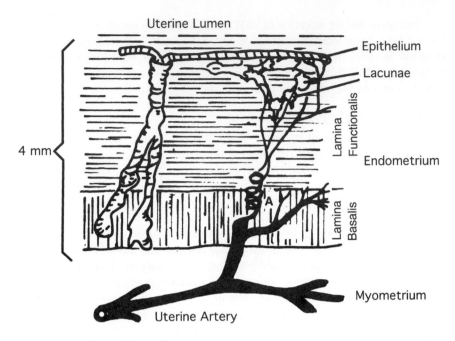

Figure 25-3 The uterine wall.
Following menstruation, the endometrium is rebuilt from the lamina basalis (proliferative phase). Spiral arteries (A) invade the developing lamina functionalis. During the secretory phase, spiral arteries and endometrial glands become highly coiled.

If implantation occurs, placental chorionic cells derived from the developing embryo produce a hormone which maintains the corpus luteum in the eventual absence of LH.

Human chorionic gonadotropin (hCG) stimulates the continued synthesis and secretion of progesterone by the corpus luteum (corpus luteum of pregnancy). hCG maintains the endometrium in the secretory phase and prevents menses. If fertilization and implantation do not occur, LH maintains the corpus luteum for about 14 days until its concentration in the blood falls below a critical level and the corpus luteum regresses. Remnants of the corpus luteum contribute to the *interstitial glands* of the ovary, which secrete small amounts of *androgens*. These androgens plus those produced by the zona reticularis of the adrenal gland are precursors of estradiol synthesized in the placenta.

With regression of the corpus luteum, maintenance of the endometrium in the secretory phase by progesterone diminishes and menses occurs.

The endometrium now enters the *menstrual phase*, where constriction of the coiled arteries in the s. functionalis leads to local ischemia, resulting in necrotic cell death and eventual sloughing of the surface epithelium and terminal portions of the coiled arteries (*menses*).

THE VAGINA AND CERVIX

The vagina serves as a site for sperm deposition, whereas the cervix presents a potential barrier between the external environment and the uterine-peritoneal cavities.

The vagina is lined by a partially keratinized stratified squamous epithelium. This mucosa is supported by a muscular coat consisting of many longitudinal bundles of smooth muscle, a few circular layers near the epithelium, and an adventitia rich in elastic tissue. There are no glands in the vagina, but a rich subepithelial capillary plexus contributes an exudate which aids in lubrication. The mucosal epithelial cells store and secrete glycogen which is broken down into lactic acid in the lumen by local lactobacilli, resulting in an acidic environment which protects against invasion by other microorganisms such as *Candida albicans*.

The increase in vaginal acidity occuring at puberty causes histologic changes in the cervical mucosa.

At puberty, hormonal changes result in eversion of the columar epithelium lining the lumen of the cervix. Exposure to the acidic milieu of the vagina leads to squamous metaplasia at the junction between the two epithelia (transformation zone). This may result in obstruction or obliteration of the cervical glands near the external opening (os) of the cervix leading to accumulation of mucus as Nabothian cysts. Repeated changes in the epithelium may also result in carcinoma of the cervix.

The presence of loose cells in the vaginal lumen is used as a diagnostic tool (exfoliative cytology).

Desquamated cells from the vagina and cervix are commonly found in the vagina and may be collected for histologic analysis (Pap smear). Cytology varies with pregnancy, since basophilic cells are indicative of the presence of both estrogen and progesterone. There is also variation with the phases of the menstrual cycle. Pap smears are useful in the early detection of cervical carcinoma.

MAMMARY GLANDS

The mammary glands are modified apocrine sweat glands.

The adult female breasts contain fat deposits and abundant connective tissue in addition to the mammary glands. The mammary glands may develop at sites anywhere along the *milk ridges* (extending from the axilla to the groin on each side), but generally appear only in the thoracic region in humans. The glands become fully developed only in postpubertal females in response to estrogens, progesterone, prolactin, and other hormones.

Growth of the mammary glands at puberty and during pregnancy is regulated primarily by reproductive hormones.

The resting mammary gland consists primarily of ductal cells divided between 15 to 20 lobes, each drained by a *lactiferous duct* at the nipple. With the onset of puberty, *estrogens* produced by the ovary stimulate cell proliferation and growth of the ducts with the addition of new *intralobular ducts*. The termini of the intralobular ducts are surrounded by *myoepithelial cells*, and the lobes are separated from each other by loose CT. Estrogens also stimulate the accumulation of fat in the breast.

Prolonged exposure to *progesterone* during pregnancy results in the differentiation of terminal intralobular duct cells into secretory (*alveolar*) cells and the formation of functional alveoli. This process is also regulated by *human placental lactogen* (placenta), *prolactin* (anterior pituitary), *thyroid hormones*, *corticosteroids* (adrenal cortex), and *local epithelial-mesenchymal interactions*.

Synthesis and secretion of milk components is regulated by prolactin.

During pregnancy, estrogen and progesterone inhibit prolactin activity. However, following birth, estrogen and progesterone levels diminish and milk production is stimulated. The first "milk" produced is *colostrum*, a watery lipid-free liquid containing a high concentration of immunoglobulins (secretory IgAs) which are transferred to the alveolar cells from plasma cells in the breast. This represents the transfer of *passive immunity* to the newborn.

> Milk is formed by a combination of merocrine and apocrine secretion.

Milk proteins are secreted by the *merocrine* mechanism. The proteins are synthesized in the rough endoplasmic reticulum (RER), packaged into vesicles in the Golgi apparatus, and the contents of the vesicles are released at the cell membrane by *exocytosis* with conservation of membrane (see Chap. 6). Nonmembrane-bound lipid droplets are stored in the apical cytoplasm of the alveolar cells. When released, they acquire a coating of membrane from the apical portion of the cell (*apocrine secretion*). Suckling by the infant initiates the *suckling or milk ejection reflex*, which is mediated by release of *oxytocin* from the posterior pituitary. Oxytocin induces contraction of the myoepithelial cells surrounding the alveoli, forcing the milk secretions into the lactiferous ducts.

> After lactation, the mammary gland regresses to the prepregnant state.

During regression, the alveolar cells are sloughed and only the myoepithelial cells, a few alveoli, and the alveolar basal lamina remain in the parenchyma. The cellular debris is phagocytosed by macrophages. During subsequent pregnancies, the alveoli will be repopulated from ductal stem cells.

With aging, the breasts undergo *involution*, where the epithelium undergoes progressive atrophy, the duct system degenerates, and the connective tissue becomes less cellular. This process includes degradation of the alveolar and ductal basal laminae, which may play a role in the development and metastasis of breast tumors. Involution of the breasts is related to the decreasing levels of estrogen and progesterone occuring at *menopause*.

> Mammary gland tumors occur in approximately one of nine North American females and are primarily of ductal origin.

Mammary gland tumors are *clonal*, i.e., they arise from one mutated cell which undergoes repeated, uncontrolled mitosis. Mammary gland tumors have been associated with mutations in a number of specific genes or gene products including p53 (a sequence-specific inhibitory gene regulatory product which inhibits the transcription of cell proliferation genes (Chap. 10 "Cell Cycle"), BRCA1 and BRCA2 (transcription factors), and erbB2 (a member of the EGF receptor superfamily; Chap. 9 "Cell Receptors and Intercellular Signaling"). Many of these markers are also present in ovarian tumors. Mammary gland tumors are *hormone-dependent*, that is, they are correlated with exposure to *estrogen*. Consequently, factors which reduce exposure to estrogen, such as delayed onset of menarche and early menopause, are correlated with reduced frequency of tumors. Delay in the onset of the first childbirth is positively correlated with increased tumor risk compared with nonchildbearing women. Nutritional factors which may contribute to increased tumor risk include high-fat diet and moderate alcohol intake.

MALE REPRODUCTIVE SYSTEM (OUT OF ONE, MANY)

If the female reproductive system is designed to produce one mature oocyte out of many candidates each cycle (e pluribus, unum), the male reproductive system is designed to make millions of sperm from hundreds of thousands of progenitor cells on a continuous basis. The two systems are similar, however, in that both are regulated by the anterior pituitary via FSH and LH and both produce haploid gametes.

THE TESTIS

The seminiferous tubules of the testes contain two types of cells, members of the germ cell lineage and somatic Sertoli cells which form a blood-testis barrier.

Sertoli cells are located around the periphery of the seminiferous tubules and are linked to each other by tight junctions (zonula occludentes). This linkage segregates an outer compartment containing the basal portion of the Sertoli cells plus *spermatogonia* and immature primary spermatocytes from a luminal compartment containing the remainder of the Sertoli cell in intimate contact with the maturing sperm cell lineage (*primary spermatocytes, secondary spermatocytes, and spermatids which give rise to sperm*). Since spermatogenesis begins at puberty, long after the development of self-recognition in the immune system (Chap. 19), these cells would be recognized as foreign by immune cells and destroyed were they not sequestered behind the Sertoli cell *blood-testis barrier*. Because spermatogonia were present during the development of self-recognition, their presence outside the blood-testis barrier does not present a stimulus for an immune response.

Spermatogenesis is the process by which spermatogonia proceed through two meiotic divisions to produce haploid spermatids. In similar fashion to the development of follicles in the ovary, spermatogenesis is regulated by FSH.

Spermatogonia come in two flavors, A and B. Type A are the stem cells and also come in two flavors, pale (Ap) and dark (Ad). Type Ad cells are thought to serve as stem cells and give rise to type Ap cells, which are committed to further differentiation. Type Ap cells become type B cells, which are precursors to *primary spermatocytes* (Fig. 25-4). Type B spermatogonia divide mitotically to give rise to primary spermatocytes, which enter the prophase of the first meiotic division. The first meiotic division is especially important, since genetic recombination occurs in the pachytene substage of prophase in both the male and female. (see Chap. 10).

Behind the protection of the blood-testis barrier, primary spermatocytes complete meiosis to form two *secondary (2°) spermatocytes* which exist only briefly (8 h) before undergoing the second meiotic division to produce four haploid *spermatids*. The intricate details of meiosis are reviewed in Chap. 10.

During spermatogenesis, primary and secondary spermatocytes and the spermatids are enclosed within folds of the Sertoli cell (Fig. 25-5). All mitotic and meiotic divisions undergo incomplete cytokinesis, thus the cells remain linked to each other by cytoplasmic bridges which ensure that sperm are genetically haploid, but phenotypically diploid.

Sertoli cells perform a myriad of important functions during spermatogenesis, which are facilitated by their close association with the developing sperm cell lineage (Fig. 25-5):

1. They provide nutritional support to these cells which are isolated from the blood supply.
2. Sertoli cells secrete a number of substances including fluid for sperm transport; *androgen-binding protein*, which facilitates the concentration of testosterone in the seminiferous tubules; and *inhibin*, which inhibits FSH release from the anterior pituitary.
3. During embryologic development, Sertoli cells produce *anti-Müllerian hormone*, which causes the regression of the paramesonephric (Müllerian) ducts,

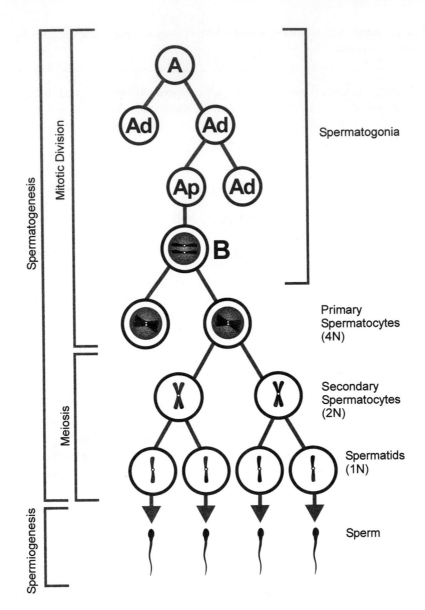

Figure 25-4 Spermatogenesis.
See Figs. 10-12 and 10-14 for a more detailed explanation of meiosis. A, A type spermatogonium; Ad, A type dark Ap, A type pale; B, B-type spermatogonium.

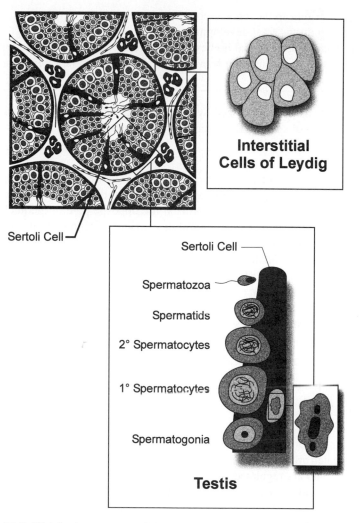

Figure 25-5 Histologic structure of the testis.
The testis is composed of seminiferous tubules. The tubules contain spermatogonia, primary spermatocytes, secondary spermatocytes, and spermatids (round and elongate) all attached to Sertoli cells. The Sertoli cells are supportive and provide a nutritive and protective function. They contain a tripartite nucleolus ("Mickey Mouse ears") within a folded or irregular nucleus (see lower insert). The spermatozoa are found free in the lumen of the seminiferous tubules. The interstitial cells of Leydig (upper insert) produce testosterone and are under the influence of the interstitial cell–stimulating hormone/lutenizing hormone (ICSH/LH) produced by the anterior pituitary.

which contribute to the female reproductive duct system (e.g. Fallopian tubes and uterus) during fetal development.

4. They phagocytose residual bodies which contain excess cytoplasm removed from spermatids during their transition to mature sperm (*spermiogenesis*).

Spermiogenesis is characterized by morphologic changes that occur during the maturation of spermatids to spermatozoa (sperm).

During the initial, or *Golgi, phase*, of spermiogenesis, the Golgi apparatus forms a cap (*acrosomal vesicle*) over the anterior portion of the spermatid nucleus, while a *flagellar axoneme* forms at the opposite pole of the nucleus. In the succeeding *acrosomal phase*, hydrolytic enzymes accumulate at high concentration in the acrosomal vesicle. These enzymes [e.g., acrosin (a serine protease), hyaluronidase and acid phosphatase] play a critical role in penetrating the outer defenses (corona radiata and zona pellucida) of the ovum during fertilization. The flagellum continues to develop during this period, and mitochondria accumulate to form the *midpiece*. During the *maturational phase*, most of the spermatid cytoplasm becomes collected into a *residual body* which is subsequently phagocytosed by the surrounding Sertoli cell.

The interstitial cells between the seminiferous tubules secrete testosterone.

The *interstitial cells (Leydig cells)* are typical steroid-secreting cells which secrete the male hormone, *testosterone*, under control of the anterior pituitary. They are located around interstitial blood vessels, are analogous to the lutein cells of the ovary, and are regulated by LH. These cells secrete androgens transiently during fetal development and then are relatively quiescent until puberty. The production of androgens during fetal development is induced by hCG. Androgens are responsible for the development of the male mesonephric (Wölffian) genital duct system.

MALE REPRODUCTIVE SYSTEM: DUCTS AND ACCESSORY GLANDS

The remainder of the male reproductive system is responsible for sperm transport, maturation, and the formation of the ejaculate.

Specific regions of the male system, their chief histologic characteristics (Fig. 25-6), and their primary functions are listed in the table on p. 358.

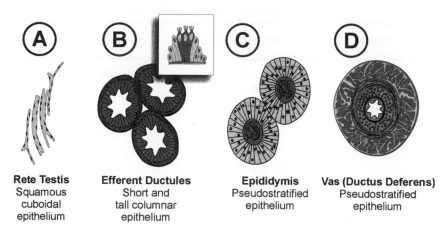

Rete Testis	**Efferent Ductules**	**Epididymis**	**Vas (Ductus Deferens)**
Squamous cuboidal epithelium	Short and tall columnar epithelium	Pseudostratified epithelium	Pseudostratified epithelium

Figure 25-6 Histologic structure of the male duct system.
The histology of the male duct system changes dramatically as when moving from the rete testis to the efferent ductules, epididymis, and eventually the vas deferens. The variation in the epithelium along the male ductal system is critical to appropriate histological identification (A–D). The vas (vas deferens, ductus deferens) has a distinctive arrangement of smooth muscle in its wall: inner longitudinal, middle circular, and outer longitudinal layer.

The prostate gland is a major site of cancer in the male.

The prostate is divided into three zones. The peripheral zone accounts for 70 percent of the gland's volume and is the origin of most prostatic tumors (adenocarcinoma), which are the second most prevalent tumors in men. At autopsy, 26 to 37 percent of men are found to have prostate cancer. The small transitional (mucosal) zone (5 percent) is the origin of most benign prostatic hypertrophy, which increases in frequency with age and may lead to obstruction of the urethra in a small percentage of cases. The central zone occupies the remainder of the gland.

FERTILIZATION

The union of ovum and sperm usually takes place in the distal third of the uterine tube (oviduct). Only a few hundred out of the several hundred million sperm present in the ejaculate penetrate this deeply into the female reproductive system.

Sperm are not initially competent to achieve fertilization.

MALE REPRODUCTIVE SYSTEM

REGION	HISTOLOGY	FUNCTION
Tubuli recti	Simple columnar epithelium, initial segment consisting of a lining of Sertoli cells	Transport of sperm
Rete testis	Simple cuboidal	Transport of sperm
Ductuli efferentes (efferent ductules)	Alternating ciliated and nonciliated cuboidal to columnar epithelium (scalloped appearance)	Transport, fluid absorption
Ductus epididymis and vas deferens	Pseudostratified columnar with stereocilia (absorptive, nonmotile)	Peristalsis by smooth muscle, sperm maturation, and storage
Seminal vesicle	Pseudostratified columnar with secretory vesicles	Produce 70% of ejaculate, including prostaglandins and sperm-activating substances such as fructose
Prostate gland	Pseudostratified to simple columnar, depending on location	Produce prostatic fluid and fibrinolysin
Bulbourethral glands	Simple cuboidal epithelium	Secrete lubricating mucus

To be able to fertilize the ovum, sperm must first undergo a process known as *capacitation*, which results from exposure to the environmental milieu of the vagina and cervix. The process of capacitation includes altering the permeability of the sperm plasma membrane covering the acrosome. This facilitates the release of acrosomal lytic enzymes required to dissolve and penetrate the zona pellucida surrounding the ovum.

Binding of sperm to the zona pellucida stimulates the *acrosomal reaction*.

The zona pellucida contains several glycoproteins including ZP1, ZP2, and ZP3. ZP2 and ZP3 are filamentous proteins cross-linked by ZP1. In addition, ZP3 possesses *O-linked oligosaccharides* which bind to a moiety (probably galactosyl transferase) on the sperm plasma membrane. Thus, ZP3 acts as a *sperm receptor*. Binding of the sperm to ZP3 results in a Ca^{2+} influx into the sperm which initiates exocytosis of the acrosomal enzymes (see Chap. 6). The requirement for specific binding of the sperm to the zona pellucida serves as a barrier to cross-species fertilization.

Once the zona pellucida is breached, binding proteins in the sperm plasma membrane initiate fusion with the plasma membrane of the ovum.

The fusion-promoting proteins (fusigens, Chap. 6) bind to integrin-like receptors on the ovum plasma membrane. Fusion results in depolarization of the oocyte plasma membrane which is a primary barrier to the binding of additional sperm (*polyspermy*). Binding of the *fusion proteins* also results in activation of a G protein (Gq) which stimulates phospholipase C to convert PIP_2 to IP_3. In turn, IP_3 stimulates release of Ca^{2+} from the smooth ER (see Chap. 9). The increase in intracellular Ca^{2+} stimulates the exocytosis of secretory granules (*cortical granules*) stored immediately below the plasma membrane of the ovum. Enzymes present in the cortical granules mediate biochemical changes (*cortical reaction*) in the zona pellucida (proteolytic cleavage of ZP2 and hydrolysis of oligosaccharides on ZP3), which also prevents polyspermy.

Following fusion of the oocyte and sperm membranes, the male pronucleus is pulled into the oocyte.

Fertilization results in the following:

1. Completion of the second meiotic division by the oocyte and extrusion of the second polar body
2. Formation of a diploid zygote
3. Determination of the sex of the embryo
4. Initiation of cleavage (mitotic division)

PLACENTA

Once fertilization has occured, the zygote is propelled from the distal uterine tube toward the body of the uterus by the cilia of the epithelium lining the uterine tube. During this process, the zygote undergoes several mitotic divisions resulting in a sphere of cells (blastocyst) with a central cavity. The cells on the outer surface of the blastocyst form the outer cell mass (trophoblast) and give rise to the placenta, whereas a group of cells accumulating at the implanting pole on the interior of the blastocyst form the inner cell mass (embryoblast) and give rise to the embryo. The process is timed so that the embryo begins implantation in the late secretory phase and is completely embedded in the uterine wall by day 27 of the menstrual cycle.

At the onset of implantation, the trophoblast divides into two layers, the outer *syncytiotrophoblast* and the inner *cytotrophoblast*.

Cells of the trophoblast divide and give rise to cells which join the syncytiotrophoblast, lose their lateral boundaries, and form a syncytium. Fingerlike processes (villi) from the syncytiotrophoblast insinuate themselves between the endometrial cells of the uterine wall and probe deeply into the underlying stroma in search of maternal capillaries. Enzymes secreted by the syncytial processes erode the capillary wall, resulting in the formation of a blood lake or lacuna.

The syncytiotrophoblast is in direct contact with maternal blood.

The villi now consist of an outer layer of syncytiotrophoblast in contact with maternal blood, an inner layer of cytotrophoblast cells, and a central core of mesenchyme which is invaded by capillaries connected to the embryonic circulatory system. Thus, nutrients, gases, and wastes must travel across the syncytiotrophoblast, the cells of the cytotrophoblast, and the embryonic capillary endothelium. Much of this transport is accomplished by a large population of pincytotic vesicles involved in transcellular transport. The cytotrophoblast layer diminishes during late pregnancy, leaving only a thin layer of syncytiotrophoblast separating maternal blood from the fetal circulation.

The maternal portion of the placenta is derived from the endometrial stroma.

The connective tissue cells of the endometrial stroma enlarge to become decidual cells. Those immediately deep to the embryo form the large decidua basalis (chorion frondosum) of the placenta, whereas the stroma in the rest of the uterine wall becomes the decidua parietalis and the stromal cells surrounding the embryo opposite the decidua basalis form the decidua capsularis, or chorion laeve. Decidual cells have prominent mitochondria and abundant RER and secrete prolactin.

Cells of the syncytiotrophoblast secrete a number of factors important for embryonic development.

In addition to numerous pinocytotic vesicles involved in maternal-fetal exchange, cells of the syncytiotrophoblast contain abundant quantities of both RER and SER, as well as lipid-cholesterol droplets, indicative of numerous secretory products including steroids. Among these are human chorionic gonadotropin (hCG), human chorionic somatomammotropin (hCS, human placental lactogen), chorionic thyrotropin and corticotropin, estrogens, and progesterone.

NERVOUS SYSTEM

•

- **Central Nervous System**
- **Meninges**
- **Cerebrospinal Fluid**
- **Blood–Brain Barrier**
- **Peripheral Nervous System**
- **Peripheral Nerve**

• • • • • • • • • • • •

The neural tissues discussed in Chap. 16 are organized into the peripheral and central nervous systems. The central nervous system (CNS) consists of the brain and spinal cord, whereas the remaining nerves and associated structures constitute the peripheral nervous system (PNS). The generalization is often made that during development the CNS forms from neuroectoderm and the peripheral nervous system forms from the neural crest. This, of course, ignores the exception that the peripheral motor system develops from the neural tube. The autonomic nervous system regulates visceral function that is nonvoluntary activity, including glandular function and modulation of cardiac and gastrointestinal function. The autonomic nervous system has two parts, the sympathetic and parasympathetic nervous systems. These two divisions often work in a "yin and yang" fashion.

CENTRAL NERVOUS SYSTEM

The common feature of the CNS is the presence of two geographic regions, white matter and gray matter.

The white matter contains myelinated axons and associated oligodendrocytes. The gray matter represents regions where neuronal cell bodies predominate along with unmyelinated axons, dendrites, synapses, and glial cells. *It is impor-tant to make the associations of nerve fibers with white matter and cell bodies with gray matter.*

The positioning of the white and gray matter is opposite in the spinal cord and the cortical areas of the brain—cerebral cortex and cerebellar cortex.

In the spinal cord there are two central horns of gray matter surrounded by the white matter (Fig. 26-1).

The positioning of the white matter and gray matter is reversed in the cerebral cortex and cerebellar cortex compared with the spinal cord. This occurs because of an additional developmental process that establishes peripheral regions of gray matter (cortex) with the development of the white matter centrally.

The cerebral cortex includes both sensory and motor areas and consists of up to six sometimes poorly defined layers within the gray matter. This gray matter is located peripherally compared with the centrally placed white matter (Fig. 26-2).

The cerebellar cortex consists of three layers: the smooth-staining molecu-lar layer, the single row of Purkinje cells forming the Purkinje cell layer, and the internal granular layer. On the inside of these layers is the white matter of the cerebellum (Fig. 26-3).

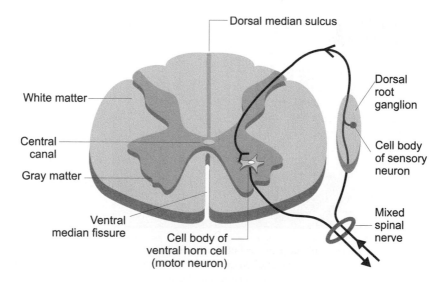

Figure 26-1 Microscopic appearance of the spinal cord.
Note the position of the gray matter on the inside (in the H-shaped horns).

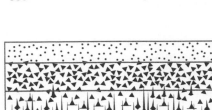
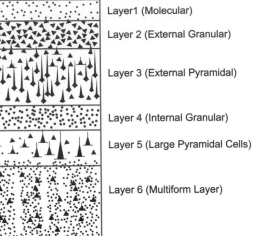

Layer1 (Molecular)

Layer 2 (External Granular)

Layer 3 (External Pyramidal)

Layer 4 (Internal Granular)

Layer 5 (Large Pyramidal Cells)

Layer 6 (Multiform Layer)

Figure 26-2 Microscopic appearance of the sensory (cerebral) cortex.
There are up to six layers of gray matter [layers I to VI starting with the clear (molecular) layer I closest to the surface]. The white matter is on the inside (not shown in the figure) in contradistinction to the spinal cord.

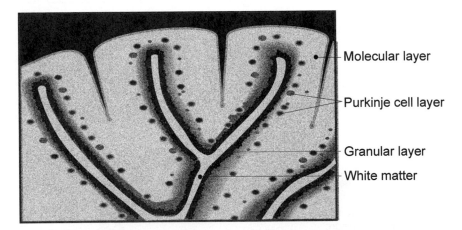

Molecular layer

Purkinje cell layer

Granular layer

White matter

Figure 26-3 Microscopic appearance of the cerebellum.
Note the position of the white matter on the inside and the cortex (gray matter) on the outside. The gray matter consists of three layers: molecular, Purkinje, and granular layers.

In addition to the cortical areas of gray matter, neuronal cell bodies are grouped as *nuclei* (not the same nucleus as the CEO of the cell). Nuclei are dispersed throughout the white matter of the brain, but in the peripheral nervous system there are no nuclei. Groupings of cell bodies in the peripheral nervous system are localized in *ganglia*.

MENINGES

The meninges form the protective layers of the brain and spinal cord.

There are three distinct layers of meninges and associated spaces moving from the outside toward the inside: (1) dura mater, (2) arachnoid, and (3) pia mater.

The *dura mater* is composed of dense connective tissue and adheres to the bone of the calvarium. The dura mater is actually composed of two layers surrounding the brain. The outer, vascular layer adheres to the bone of the cranium surrounding the brain and is therefore called the *endosteal layer*. This layer is important because it functions as the periosteum in that region. It retains osteogenic potential throughout life and therefore is a source of osteoblasts during fracture healing. The inner layer of the dura is called the *fibrous layer*, and its inner surface is covered by a single layer of cells derived from mesenchyme. The dura of the spinal cord consists only of a fibrous layer, and the bones of the vertebral column have their own periosteum. The *arachnoid* is a weblike structure sandwiched between the dura and the vascular *pia mater*, which is found closest to the brain or spinal cord. There are a number of spaces between the meninges: (1) the *epidural* (between the periosteum and the dura), (2) *subdural* (between the dura and the arachnoid), and (3) the *subarachoid* (between the arachnoid and the pia).

The epidural space is used as an alternative route for anesthetic in place of general anesthesia and is used particularly during childbirth. The subdural space functions as a fluid-filled hydraulic cushion which protects the CNS from traumatic injury. The subarachnoid space is the location of the cerebrospinal fluid.

CEREBROSPINAL FLUID

The cerebrospinal fluid (CSF) is produced by the choroid plexus. The CSF is a clear, viscous fluid which contains virtually no protein and is in equilibrium with the extracellular brain fluid.

The CSF functions in shock absorption, volume-transmission, and metabolic regulation.

The CSF provides a fluid-filled cushion for the CNS and spreads growth factors and cytokines from a localized area to entire regions of the CNS. It is an ultrafiltrate of the plasma, but differs from the plasma in certain key constituents. The water content of the CSF is higher and protein content very much lower [35 mg/dL (deciliters) compared with 7000 mg/dL in serum]. Ionic concentrations also vary between CSF and serum with K^+ and Ca^{2+} lower, whereas Mg^{2+} and Cl^- are higher in the CSF. The pH is lower in the CSF. All these factors affect the excitability of neural (brain) tissue.

The CSF circulates through the foramina of Magendie and Luschka that allow drainage from the ventricles of the brain into the subarachnoid space. Absorption of the CSF occurs through the venous sinuses in the dura mater.

BLOOD–BRAIN BARRIER

The blood-brain barrier prevents the flow of large molecules from the blood to the brain. This barrier represents a blood-CSF barrier since the CSF is in equilibrium with the extracellular fluid of the brain.

The blood-brain barrier consists of a continuous capillary endothelium (no fenestrations) with occluding junctions (zonula occludentes, tight junctions) between endothelial cells in brain capillaries.

The blood-brain barrier is induced by astrocytic foot processes that surround the capillaries, but the foot processes do not participate directly in the barrier function. The astrocytes contribute to the blood-brain barrier by secretion of a factor which stimulates tight junction formation between endothelial cells. The paucity of pinocytotic vesicles within the brain capillary endothelial cells also contributes to the barrier, but the tight junctions provide the major obstacle to molecules. The solute characteristics also determine permeability of the blood-brain barrier. The barrier generally does not permit protein passage, explaining the low content of albumin in the CSF. The barrier is more permeable to smaller and more lipid soluble molecules; ionic polar molecules are poorly permeable unless a transport system is operational. There are specific carrier-mediated transport systems associated with the blood-brain barrier. These transport systems include specific amino acids, sugars, and other metabolites. A barrier is also present around nerves (blood-nerve barrier).

PERIPHERAL NERVOUS SYSTEM

The peripheral nervous system contains nerves, nerve endings (sensory receptors), and aggregations of perikarya called ganglia.

PERIPHERAL NERVE

A neuron in the peripheral nervous system consists of a single axon and multiple dendrites with a perikaryon or cell body containing the nucleus. Peripheral nerves consist of afferent (sensory) and efferent (motor) fibers bundled with connective tissue.

The endoneurium, perineurium, and epineurium are connective tissue ensheathments that can be observed when moving from the axon toward the periphery. The convention of epi-, peri-, and endo- is identical with the nomenclature for muscle epimysium, perimysium, and endomysium (Chap. 15, "Muscle").

The epineurium encircles the entire nerve. Bundles of nerve fibers are surrounded by a sheath called the perineurium. Although the perineurium is formed from connective tissue, it is epithelial-like, contractile, and contains tight junctions. The perineurium is the primary macromolecular barrier, but the other connective tissue sheaths may also contribute to the maintenance of the excitable environment of the nerve fiber (blood-nerve barrier). Individual nerve fibers are surrounded by connective tissue called the endoneurium. Closest to the nerve fiber is the Schwann cell and its sheath. The Schwann cell is derived from the neural crest and is responsible for myelination in the peripheral nervous sytem (Fig. 26-4).

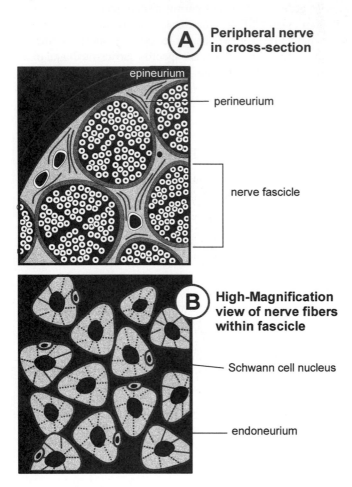

Figure 26-4 Peripheral nerve and myelin.
A. Two of the three wrappings of a peripheral nerve are shown (epineurium and peri-nurium). Note the similarity of the prefixes endo-, peri-, and epi-, which were also used in the nomenclature for the connective tissue wrappings of muscle fibers. *B.* The appearance of individual nerve fibers at high magnification. The dashed lines indicate the washed out appearance that occurs after processing through alcohols and organic solvents. The removal of myelin leaves a "wagon wheel" appearance in the region surrounding the centrally placed axon (dark central circle). Schwann cell nuclei and the endoneurium are also depicted.

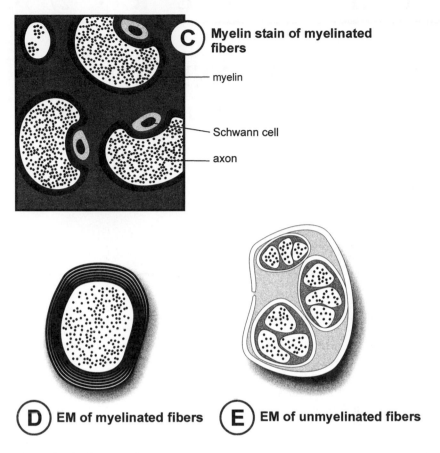

C. Myelin stain of myelinated fibers

myelin

Schwann cell

axon

D. EM of myelinated fibers   E. EM of unmyelinated fibers

Figure 26-4 *(Continued)*
C. The appearance of nerve stained with osmium (myelin) stain. *D.* The electron microscopic appearance of a myelinated fiber (1 : 1 ratio between Schwann cell and axon segment). *E.* The electron microscopic appearance of unmyelinated fibers (more than one
fiber/Schwann cell). Black dots within the neuronal processes represent microtubules.

SIGNAL TRANSDUCTION: THE SPECIAL SENSES

·

- **Eye**
- **Ear**
- **Olfaction**
- **Taste**

·   ·   ·   ·   ·   ·   ·   ·   ·   ·   ·   ·

The special senses (vision, hearing, taste, and olfaction) keep humans in touch with the dangers as well as the aesthetic aspects of the environment, such as the Mona Lisa, a Hootie and the Blowfish concert, a bowl of New Orleans gumbo, and coffee brewing in the morning. One thing they share in common is the ability to transduce a chemical (taste and olfaction) or physical stimulus (vision and hearing) into electrical impulses to the brain. The systems are, however, dissimilar in significant ways. In the olfactory system, for instance, the cell receiving the stimulus is the afferent neuron itself, whereas in the other systems one or more specialized cells (receptor cells and interneurons) are interposed between the stimulus and the afferent neuron. The ensuing pages concentrate first on the general histology and then on the histophysiology and cell biology most intimately related to signal transduction for each sense organ.

EYE

The eye consists of three layers (sclera and cornea, uvea, and retina) and three compartments (anterior chamber, posterior chamber, and the vitreous space).

The *sclera* (Fig. 27-1), or outer fibrous coat of the eye (tunica fibrosa), consists of numerous collagen bundles in various orientations with regard to each

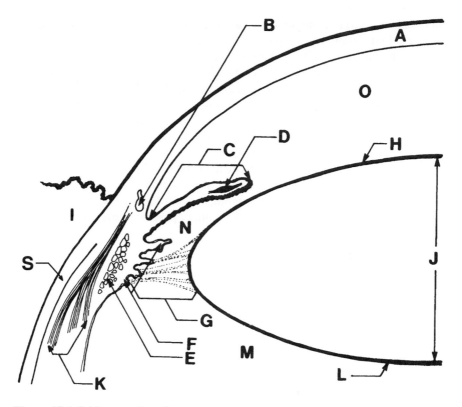

Figure 27-1 Iridocorneal angle.
A. Cornea. B. Canal of Schlemm. C. Iris. D. Sphincter pupillae muscle. E. Ciliary muscle.
F. Ciliary processes. G. Zonule fibers. H. Lens epithelium. I. Conjunctiva. J. Lens. K. Ciliary muscle. L. Lens epithelium. M. Vitreous body. N. Posterior chamber. O. Anterior chamber. S. Sclera.

other but always parallel to the outer surface of the sclera. Posteriorly, the sclera is continuous with the dural covering of the optic nerve. Anteriorly, the sclera is continuous with the *cornea*. The sclera possesses relatively few blood vessels, whereas the cornea is completely avascular. However, the junction between the two, the *limbus*, is highly vascularized. The cornea is covered on the outer surface by a stratified, nonkeratinized squamous epithelium of five to six layers which, owing to high mitotic activity in the basal layer, is constantly renewed with a turnover period of about 1 week. An acellular zone of collagen fibers (*Bowman's membrane*) separates the corneal epithelium from the underlying corneal *stroma*, which consists of flattened fibroblasts and highly ordered arrays of parallel collagen bundles and ground substance. Another homogeneous layer of collagen bundles (Descemet's membrane) separates the stroma from the endothelium, a layer of simple squamous cells. Corneal epithelial and endothelial cells help to maintain the transparency of the corneum (see below).

The middle layer, or *uvea*, consists of the *choroid*, the *ciliary body*, and the *iris*, and is well vascularized (Fig. 27-1). The choroid contains numerous collagen and elastic fibers, fibroblasts, lymphocytes, and other immune cells as well as melanocytes. The *retina* (see below) is metabolically maintained by a very vascular, inner layer of the choroid, the *choriocapillary layer* (choriocapillaris). Anteriorly, the choroid is expanded to form the ciliary body. On the surface facing the interior of the eye, the ciliary body is covered by a double layer of cells, the innermost a continuation of the pigmented layer of the retina and the surface cells being derived from the sensory layer of the retina. *Ciliary muscles* within the body alter the tension on the lens through *zonule fibers* (oxytalan fibers; fibrillin, Chap. 13) projecting from fingerlike ciliary processes and inserting on the lens capsule.

The ciliary muscles change the shape of the lens (accommodation).

When the eye focuses on objects in the distance, one group of ciliary muscles stretches the choroid, keeping traction on the zonule fibers and flattening the lens. Contraction of the other group of ciliary muscles draws the ciliary body anteriorly, reducing the tension on the zonule fibers and allowing the lens to thicken for focusing on near objects. This is the process of *accommodation*.

A long radial process of the ciliary body, the iris, regulates the amount of light entering the eye.

The iris surrounds a central opening, the *pupil* (Fig. 27-1). A double-layered epithelium is found on the posterior surface of the iris. The inner layer consists of pigmented epithelial cells, whereas the outer layer consists of specialized epithelial cells (*dilator pupillae muscle*) with radially oriented myofilaments (myosin and actin). Contraction of these cells in response to sympathetic innervation results in dilation of the pupil and more light entering the eye. A group of smooth muscle cells, the *sphincter pupillae*, is arranged concentrically in the leading edge of the iris and contracts the pupil under parasympathetic innervation. Melanocytes of the pupil are responsible for the color of the pupil: the presence of only a few melanocytes results in blue eyes, whereas larger numbers result in darker shades.

> The nonpigmented, inner (surface) layer of the epithelial cells of the ciliary body produce aqueous humor.

These cells have basal infoldings characteristic of ion-transporting cells and produce a protein-poor filtrate of blood plasma (*aqueous humor*) which is secreted into the *posterior chamber*, the space between the lens and the iris. Aqueous humor reaches the *anterior chamber* of the eye between the iris and the cornea through the pupil (Fig. 27-1). At the margins of the cornea, the aqueous humor percolates through a *trabecular meshwork* (spaces of Fontana) to reach the *canal of Schlemm*, an irregular sinus lined with endothelium and connected to small scleral veins. *Glaucoma* results from the increase in intraocular pressure occuring when blockage of the outflow passages prevents the removal of excess aqueous humor.

The vitreous compartment posterior to the lens is occupied by a transparent gelatinous substance, the *vitreous humor*, consisting mostly of water, fine collagen fibers, and a high concentration of hyaluronic acid.

> The structures of the eye through which light must pass to reach the rods and cones (light receptors) are specialized to be transparent.

To reach the receptor cells, light must pass through the *cornea*, the anterior chamber filled with aqueous humor, the *lens*, the posterior chamber filled with vitreous humor, and the neural layers of the retina (Fig. 27-1). To prevent the diffraction of light and a blurry image, the components of these structures must be as "transparent" as possible. The cells forming these structures contain relatively few intracellular organelles such as mitochondria which would scatter photons. Another source of diffraction is removed by the scarcity of blood vessels in the preretinal light path, nutrients being supplied, and metabolites removed, by the aqueous humor. Like the brain, the internal structures of the eye receive most of their energy from aerobic metabolism of glucose using enzymes which are soluble in the cytoplasm. Transparency is further aided by the highly ordered orientation of structural elements such as the thin collagen bundles in the corneal stroma and the crystallin proteins of the lens.

> The lens is a biconvex disk constructed of specialized epithelial cells.

The outer capsule of the lens (Fig. 27-1) is actually a thick basement membrane secreted by a single layer of subcapsular epithelial cells and consisting primarily of glycoproteins and type IV collagen. It is highly refractile to light. The subcapsular epithelial cells are cuboidal in shape and contiguous with the cells forming the "fibers" of the lens, for which they serve as precursors. Lens cells originate in the embryo from the epithelium on the posterior surface of the lens vesicle and are present throughout life. Lens cells mature into long, thin, lifeless cells lacking nuclei and most cytoplasmic organelles. However, their population is augmented by epithelial (subcapsular) cells on the anterior surface of the lens which proliferate, are pushed to the equator of the lens, and differentiate into lens cells. Lens fibers are filled with *crystallin proteins (α, β, and γ)*, which are maintained in an unaggregated state by osmotic balance maintained by Na^+, K^+-ATPase pumps. *Cataracts* (opacities) develop in the lens when changes in lens cell osmolarity alter the solubility of the crystallin proteins. This may result from physical or chemical injury. Senile cataracts develop as the result of age-related changes in protein metabolism, whereas diabetic cataracts result from increased osmolarity due to intracellular accumulation of sorbitol caused by high lens glucose concentration. Cataracts may also form congenitally as a result of rubella exposure *in utero* or galactosemia in the neonatal period.

The sensory apparatus of the eye is the retina.

The *retina* is a multilayered structure lining the interior surface of the posterior chamber of the eyeball (Fig. 27-2). The peculiar organization of the retina with neural tissue adjacent to the lumen and an "epithelium" on the outer periphery is the result of formation from the *optic vesicle*, a double-walled outgrowth from the brain which invaginates to form the *optic cup*. (For further information on the embryologic origins of the eye, see *Basic Concepts in Embryology* by L.J. Sweeney, a companion book in this series.)

The retina consists of several layers of neuronal cells and a pigmented epithelium.

The layer adjacent to the lumen of the posterior chamber consists of *ganglion cells*, neurons whose afferent processes (axons) form the *optic nerve*. The layer adjacent to the peripherally placed *pigmented epithelium* is formed by modified neurons, the *rod and cone cells*, which are the light receptors (Fig. 27-2). Between the ganglion cell layer and the rod and cone cells are layers of interneu-

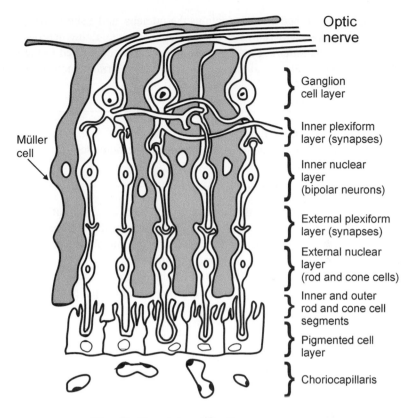

Optic
nerve

} Ganglion
cell layer

} Inner plexiform
layer (synapses)

Müller
cell

} Inner nuclear
layer
(bipolar neurons)

} External plexiform
layer (synapses)

} External nuclear
layer
(rod and cone cells)

} Inner and outer
rod and cone cell
segments

} Pigmented cell
layer

} Choriocapillaris

The Retina (simplified)

Figure 27-2 The organization of the retina.
In this simplified schematic of the retina, photoreceptors (rods and cones), bipolar neurons, and ganglion cells are shown. Müller cells are glial elements which form a structural framework.

rons (*bipolar cells, horizontal cells, and amacrine cells*), which integrate and amplify the signals from the receptor cells and relay them to the ganglion cells. Since it results from a direct outgrowth of the brain, the retina includes various types of glial cells, including *cells of Müller*, which span the neural layers.

Rod and cone cells are extensively modified neurons.

Rods and cones (Fig. 27-2) share a similar organization, a slightly different morphology, and different sensitivities to light (light intensity and color, respectively). Both cell types consist of an *inner segment* containing mitochondria and other organelles, a cell body which contains the nucleus, and a short axon which makes synaptic contact with relaying (bipolar cells) and integrating neurons (e.g., horizontal cells). The *outer segment* is a modified cilium connected to the inner segment by a short stalk containing the typical ciliary arrangement of microtubules. (See Chap. 8.)

> The photoreceptors are transmembrane glycoproteins (e.g., rhodopsin) found in stacks of membranous disks located in the outer segment of rod and cone cells.

The membranous disks of the outer segment form as invaginations of the cell membrane. In rod cells, they completely detach from the cell membrane to form free disks while they remain connected to the cell membrane in cone cells. The membranous disks and their photoreceptor proteins are constantly renewed in rod cells by formation near the base of the outer segment and extrusion from its tip where the old disks are phagocytosed by the surrounding pigmented epithelial cells (retinal pigmented epithelium, RPE).

> Rhodopsin is the photosensitive transmembrane glycoprotein of the membranous stacks in rod cells.

Rhodopsin consists of *opsin* and a prosthetic group, 11-*cis*-retinal. When a photon strikes the 11-*cis*-retinal, it undergoes a conformational change to all-*trans*-retinal and rhodopsin becomes activated when the *trans*-retinal dissociates from the complex.

> A second messenger, cyclic GMP (cGMP), couples the conformational change in rhodopsin to the closing of Na^+ channels in rod cells.

The activated rhodopsin interacts with a G protein, *transducin (G_t)*, which activates *cGMP phosphodiesterase* to hydrolyze intracellular cGMP. Since cGMP is directly responsible for maintaining *rod cell Na^+ channels* in the open state, the precipitous decline in cGMP results in the closing of the channels and a hyperpolarization of the rod cell. (Note that this is a substantially different process from the usual receptor–G_s interaction which stimulates cAMP formation through activation of adenylate cyclase.) *Amplification* throughout the cascade results in the hydrolysis of hundreds of thousands of cGMP molecules and closure of hundreds of Na^+ channels for each photon. (See Chap. 9.)

As with other receptor–G-protein complexes, the interaction is terminated by a variety of regulators.

The β/γ subunit of transducin phosphorylates the activated photoreceptor (opsin), inducing the binding of *arrestin*, which inhibits the interaction between opsin and the transducin α subunit. *Trans*-retinal is converted to 11-*cis*-retinal in the cytoplasm and rebinds to opsin. The genes for many components of the photo-transduction cascade [arrestin, *phosducin (a regulator of the β/γ subunit of transducin)*, intraphotoreceptor binding protein (IRBP), and phosphodiesterase] contain an identical regulatory sequence, the *photoreceptor conserved element 1 (PCE1)*.

Hyperpolarization of the rod cell results in the disinhibition of signal to the brain.

Under dark conditions, rod cells are in a depolarized state, which results in a constant influx of Ca^{2+} at the synaptic terminals of the inner segment and the consequent release of an inhibitory neurotransmitter, which reduces the activity of the bipolar neurons and subsequently the ganglion cells which form the optic nerves. Closing of the Na^+ channels results in hyperpolarization of the rod cell with a subsequent reduction in Ca^{2+} influx at the presynaptic terminals. Decreased release of inhibitory neurotransmitter results in increased activity of bipolar and ganglion cells.

A calcium-dependent mechanism allows photoreceptors to adapt.

Closure of the sodium channels also reduces the influx of Ca^{2+}. The fall in intracellular Ca^{2+} concentration activates *recoverin*, a mediator similar to calmodulin but with opposite sensitivity to Ca^{2+}, which stimulates guanylate cyclase to return the cGMP concentration to normal. This mechanism plays an important role in the adaptation to continuous light exposure.

Cone cells utilize the same signal transduction method as rod cells.

Cone cells, instead of rhodopsin, use at least three slightly different transmembrane glycoproteins, *iodopsins*, which are selectively sensitive to the red, green, and blue wavelengths of light.

EAR

The sensory apparatuses of the inner ear include mechanisms for transducing sound (*cochlea*) as well as angular momentum (*semicircular canals*) and position sense (*saccule and utricle*). All three modalities transduce mechanical movements into electrical signals and are dependent on specialized *hair cells*. The inner ear consists of a *bony labyrinth* of spaces within the petrous temporal bone which house the *membranous labyrinth*, a continuous epithelium-lined tube that includes the utricle and attached semicircular ducts and the saccule with attached cochlear duct. The bony labyrinth is filled with *perilymph*, whereas the membranous labyrinth contains *endolymph*. Both fluids are similar to extracellular fluid but have a low protein content. In addition, the endolymph has a higher K^+ content and a lower Na^+ content than the perilymph. The high K^+ content of the endolymph plays an important role in signal transduction.

In the cochlea, the *cochlear duct (scala media)*, flanked by the *scala vestibuli* and the *scala tympani* of the bony labyrinth (Fig. 27-3), takes 2.5 spiral

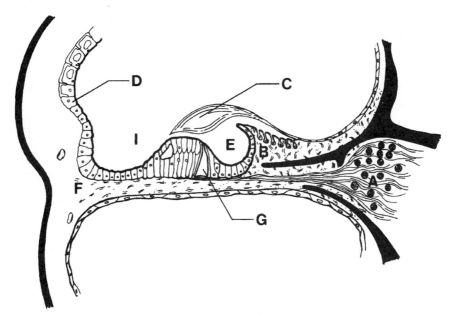

Figure 27-3 The organ of Corti.
A. Spiral ganglion. *B.* Spiral limbus. *C.* Tectorial membrane. *D.* Stria vascularis. *E.* Internal spiral tunnel. *F.* Spiral ligament. *G.* Inner tunnel separating outer and inner hair cells. *I.* Cochlear duct (scala media).

turns around a central bony core, the modiolus, which houses the spiral ganglion. The scala vestibuli and scala tympani are lined by a simple squamous epithelium and are continuous at the helicotrema, the apex of the cochlea. The scala media is separated from the scala vestibuli by a double layer of squamous epithelial cells (*vestibular membrane*) and from the scala tympani by the *basilar membrane* consisting of a simple layer of mesothelium from the scala tympani and a thick layer of extracellular matrix.

The epithelium of the *stria vascularis* (Fig. 27-3) in the lateral wall of the scala media covers a well vascularized layer of connective tissue. Epithelial cells of the stria are characterized by basal infoldings with numerous associated mitochondria, indicating that these cells are involved in ion and water transport and may regulate the specialized high K^+ composition of the endolymph. The remainder of cells lining the lateral wall and lateral portion of the basilar membrane form a single layer of cuboidal to low columnar cells.

The *organ of Corti* occupies the medial portion of the basilar membrane in the cochlear duct. The organ consists primarily of columnar supportive (*phalangeal*) cells, *pillar cells*, and hair cells, as well as the overlying *tectorial membrane*. Pillar cells are stiffened by numerous microtubules and form a structural support for the inner tunnel. Phalangeal cells support the hair cells and form tight junctions with them, sequestering the basal portion of the hair cells from the high K^+ environment of the endolymph. The three to five rows of outer hair cells are characterized by a V- or W-shaped arrangement of stereocilia on their apical surface, the tallest of which is embedded in the tectorial membrane. The single row of inner hair cells possess a linear arrangement of stereocilia which do not contact the tectorial membrane. Hair cells belong to a peculiar fraternity of cells (e.g., most neurons) that are not replaced during life—so turn down the volume on the boom box. The functional histology of sensory portions of the semicircular ducts (*ampullae*) and the utricle and saccule (*maculae*) is similar to that of the organ of Corti. In the ampullae, an elevated ridge (*crista ampullaris*) contains supportive cells and hair cells (Fig. 27-4). The stereocilia of the hair cells are embedded in an overlying gelatinous membrane, the *cupula*. In the maculae of the utricle and saccule, the hair cells possess several rows of stereocilia and a single kinocilium (a true cilium with an axoneme) embedded in the gelatinous matrix of the *otolithic membrane* (named for the presence of small granules [otoliths] consisting of protein and calcium carbonate).

Hair cells of the cochlea (organ of Corti) transduce sound in the form of fluid pressure waves.

Actually, two mechanical transduction events are required for hearing. The first occurs when pressure waves in the air vibrate the *tympanic membrane*

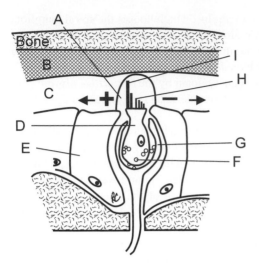

Figure 27-4 Schematic drawing of a typical hair cell of the cristae ampullaris.
Bending of the kinocilium in the plus direction increases the response of the hair cell, whereas bending in the minus direction decreases the activity of the hair cell. *A.* Cupula. *B.* Bony labyrinth. *C.* Membranous labyrinth. *D.* Hair cell. *E.* Supporting cell. *F.* Neurotransmitter vesicle. *G.* Chalice-shaped terminal of a ganglion cell. *H.* Stereocilia. *I.* Kinocilium. *(Courtesy of Dr. Robert J. Cowie.)*

(eardrum). The ossicles of the middle ear *(malleus, incus, and stapes)* transmit these vibrations to the fluid-filled inner ear via the oval window.

> Hair cells of the organ of Corti possess specialized apical microvilli (stereocilia) embedded in a gelatinous (tectorial) membrane.

Hair cells form a portion of the epithelial lining of the cochlear duct situated between the vestibular and tympanic cavities. Fluid pressure waves in the vestibular cavity result in movement of the vestibular and basilar membranes which cause deflections in the hair cell *stereocilia* relative to the stable *tectorial membrane.*

> The stereocilia of the hair cells are mechanoreceptors linked directly to membrane ion channels.

The stereocilia of hair cells are arranged in parallel rows of decreasing height. When pressure waves in the cochlea produce a shearing movement of the hair cells with respect to the tectorial membrane, the stereocilia become deflected in unison in one direction. The tip of each stereocilium is connected by *microfilaments* to the tips of adjoining stereocilia in such a way that deflection of the longest stereocilium tugs the others along with it. Current evidence strongly suggests that the microfilaments are physically connected to *ion channels* in the tips of the stereocilia such that mechanical deflection opens the channels like trap doors (mechanically gated ion channels).

Opening of the ion channels results in an inward flux of K^+.

The inward flux of K^+ depolarizes the membrane potential of hair cells, resulting in the release of stimulatory neurotransmitter at the synapse with a ganglion cell. *Recall that the endolymph of the cochlear duct is rich in K^+.*

The mechanical transduction of signal is the same for hair cells in the ampullae of the semicircular canals and the maculae of the utricle and saccule.

In the *ampullae* of the semicircular canals, movements of the cupula in the fluid-filled semicircular canals result from changes in angular acceleration (i.e., in the direction and amplitude of movement in the x, y, and z planes) which cause deflections of the embedded stereocilia. As in the cochlea, this results in opening of membrane ion channels and subsequent depolarization of the cell membrane.

In the *maculae* of the utricle and saccule, changes in the position of the head with respect to the gravitational field result in deflection of the hair cell stereocilia and mechanical opening of their ion channels.

OLFACTION

The cells responding to odorants (chemical signals) are *bipolar neurons* localized to special regions of the respiratory system (the olfactory mucosa and the vomeronasal organs). Although cells of the former are sensitive to a variety of odorants, the cells of the latter appear specialized for the detection of pheromones (agents which mediate sexual behavior in mammals).

The olfactory receptors are unique among the special sense organs. The receptor cells are the afferent neurons.

The olfactory receptor cells are modified neurons, the principal modification being the presence of several cilia on their dendritic ends which protrude into the airway lumen (Fig. 27-5). The cilia are large but probably only minimally motile, since a typical axoneme (Chap. 8) is only present in the proximal portion of each cilium. The receptor cells are surrounded by supporting (*sustentacular*) cells and basal *stem cells*. The stem cells give rise not only to supporting cells but also to new receptor cells (1 month lifespan). *Thus, the receptor cells represent one of the very few constantly regenerating populations of neurons in the body.*

> The surface cilia contain numerous transmembrane receptors for odorants.

There may be hundreds of different receptor proteins, each responsive to a specific chemical odorant or closely related odorant family. The receptors are

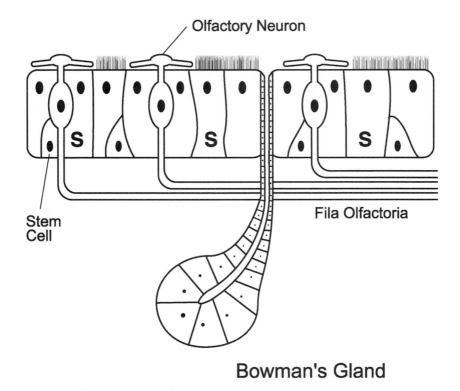

Figure 27-5 The olfactory epithelium.
Specialized cilia on the dendritic end of olfactory neurons contain G protein–linked receptors and protrude into the airway lumen. Glandular secretions (serous) likely aid in solubilizing odorants. S = sustentaeular cells.

typical seven-pass transmembrane proteins which interact with *olfactory G proteins (G_{olf})* to stimulate adenylate cyclase and increase intracellular *cAMP* content. In these cells, cAMP binds directly to, and opens, Na^+ channels, resulting in Na^+ influx and depolarization of the cell. When depolarization reaches threshold, an action potential is generated which travels along the axon (*fila olfactoria*) to the *olfactory bulb* of the brain.

TASTE

The sensation of taste is restricted to four elements: sweet, acidic (sour), bitter, and salty (not jalapeno). Further information is provided by the olfactory system. For example, a a test subject, blindfolded with his nostrils plugged, could not tell the difference between an apple and an onion by taste alone. Similarly, all people experience a deficit in the ability to taste foods when they suffer from the plugged nasal passages accompanying a cold.

The sensory organs for taste, the taste buds, are associated with the foliate, fungiform, and circumvallate papillae on the dorsal surface of the tongue.

Taste buds consist of a barrel-stave arrangement of tall supporting cells and receptor cells organized around a central pore (also see tongue section of Chap. 22). Stem cells are present at the base of the taste bud. Taste receptor cells are characterized by the presence of concentrations of synaptic-like vesicles adjacent to the location of postsynaptic dendritic terminals of small neurons on the receptor cell surface. In addition to the presence of synaptic vesicles, the identity of receptor cells as modified neurons is further suggested by the presence of a neuronal marker, *neuron-specific enolase (NSE)* in subpopulations of these cells.

Taste receptors are located primarily on the apical microvilli of the receptor cells and are associated with a G protein.

One G protein associated with taste receptors (bitter-sweet) is *gustducin (G_g)*. It is very similar (80 to 90 percent amino acid sequence homology) to transducin, the G protein associated with photoreceptors. Evidence suggests that signal transduction occurs in a similar process to that of the retinal receptors. Activation of a *cAMP* phosphodiesterase results in reduction of intracellular cAMP concentration and closing of membrane Na^+ channels with subsequent hyperpolarization of the cell leading to disinhibition of the afferent neurons. Since the sweet response is dependent on increased cAMP, activation of gustducin could either inhibit the sweet response or produce the bitter response. The salty and sour (acid) responses are the result of direct effects on ion channels, the former resulting in Na^+ influx and the latter in H^+ blockade of K^+ channels.

Signal transduction in the special senses is summarized in the table.

SIGNAL TRANSDUCTION

MODALITY	STIMULUS	RECEPTOR	G PROTEIN	SECOND MESSENGER	ION CHANNEL
Olfaction	Chemical	Yes	G_{olf}	Increased cAMP	Open Na^+
Taste	Chemical (bitter)	Yes	G_g	Decreased cAMP	Close Na^+
Hearing and position	Mechanical	No	—	—	Open K^+
Vision	Physical (photons)	Yes	G_t	Decreased cGMP	Close Na^+

THE STRATEGY

·

- **Decision Tree of Organ Identification**
- **Strategies for the Nervous System**
- **Bone, Cartilage, and Muscle Strategies**
- **Blood Vessel Strategy**
- **Endocrine Glands Strategy**
- **Exocrine Glands Strategy**
- **Female Reproductive System Strategy**
- **GI Tract Strategy**
- **Lymphoid Organs Strategy**
- **Male Reproductive System Strategy**
- **Respiratory System Strategy**
- **Urinary System Strategy**

·   ·   ·   ·   ·   ·   ·   ·   ·   ·   ·   ·

There is no mystery to histology. For the most part (99 percent of the time), cells, tissues, and organs can be identified using a straightforward set of criteria. The student's task is to learn this set of criteria (rules) for diagnostic identification of tissues and organs. The criteria have been formulated and validated by the Scientific Method of (1) observation, (2) data analysis, (3) hypothesis formation, and (4) hypothesis testing. People who have to discover the rules for themselves probably take more time than they expect to spend, but it is guaranteed that those who work them out for themselves never forget them and emerge from the process more expert. OK, back to reality. Here is a helpful, short set of rules and strategies. It is important to complete this assignment to mentally label the visual images of histology so that they can be recalled to make an "appropriate diagnosis."

DECISION TREE OF ORGAN IDENTIFICATION (OR "THE HITCHIKER'S GUIDE TO MYSTERY MEAT")

The set of rules and strategies for organ and tissue identification (sometimes called "the hitchhiker's guide to mystery meat") as formulated by the Scientific Method, are a short set of decisions in the form of a branching tree or flow chart. The questions proceed from the very general to the more specific. Each succeeding question enables narrowing the field of possibilities. Begin with the obvious: Is the organ hollow (i.e., does it have a lumen) or is it basically a solid mass of cells? (See table).

POSSIBILITIES

HOLLOW	SOLID
Cardiovascular system	Endocrine glands
Gastrointestinal system	Exocrine glands
Urinary tract	Connective tissue
Respiratory system	Muscle
Reproductive organs	Nervous tissue
(except ovary)	Skin
	Lymphoid organs
	Ovary

OK. It is decided that it is an organ with a lumen. By definition, all lumens must be lined by an epithelium. Each of the four hollow organ systems listed in the table has a lumen lined by a very specific type of epithelium. The next question should therefore be: What type of epithelium lines the lumen of this organ? (See the table.)

POSSIBILITIES

SIMPLE SQUAMOUS	PSEUDOSTRAT. CIL. COL.	SIMPLE COL.	TRANSITIONAL
Cardiovascular	Respiratory	GI	Urinary

Now the epithelium in question is identified as simple columnar. The possibilities have now been narrowed down from all the tissues and organs in the body to one particular organ system with just two questions. Continuing with the GI system as an example, for the most part, the GI tract is lined by a simple columnar epithelium. In one part of the tract, the small intestine, the epithelium is modified to form villi. The next question is: does the epithelium have villi? (See the table.)

POSSIBILITIES

YES	NO
Duodenum	Stomach
Jejunum	Gallbladder
Ileum	Large intestine

Assume villi are seen. The next task is to distinguish between the three portions of the small intestine. What are the defining characteristics of each part? The duodenum contains Brunner's glands in the submucosa. Does the specimen contain Brunner's glands? No? Then that leaves either jejunum or ileum. How do these two segments of the small intestine differ? Peyer's patches (groups of lymphoid nodules) occur only in the ileum. If there are Peyer's patches, the specimen is from the ileum. If not, then the tissue on the slide is jejunum.

One more key to deciphering tissues and organs is to remember the exceptions to the rules. Simple columnar epithelium characterizes the GI tract. However, the ends of the GI tract, the esophagus and anus, are lined by stratified squamous epithelium. How is this distinguished from skin? Recall that the GI tract has a regular pattern of epithelium, connective tissue (lamina propria), muscle (muscularis mucosa), connective tissue (submucosa), muscle (muscularis externa), and connective tissue (serosa or adventitia). This pattern is not found in the skin. Decision trees for the major tissues and organ systems are found in Figs. 28-1 to 28-10.

THE THREE MOST IMPORTANT RULES

1. Start with the lowest possible magnification (your naked eye) and work up.
2. Never rely on color.
3. Never memorize the images seen in the microscope or atlas.

STRATEGIES FOR THE NERVOUS SYSTEM

In longitudinal section, peripheral nerves appear wavy and washed out. This is due to removal of lipid from myelin sheaths during processing in alcohol and xylene. In cross section, the appearance is that of a bundle of dark cylinders within lighter cylinders resulting in a "wagonwheel" appearance. Nuclei of fibroblasts (endoneurium), Schwann cells, and a few endothelial cells are visible within peripheral nerves. *There are no cell bodies of neurons in peripheral nerves.* In the peripheral nervous system, the neuronal cell bodies are found in the dorsal root ganglia and autonomic ganglia.

The central nervous system contains neuronal cell bodies of many sizes and shapes as well as three types of glial cells (astrocytes, oligodendroglia, and microglia). The basic architecture of the spinal cord, the cerebrum, and the cerebellum is outlined in Chap. 26.

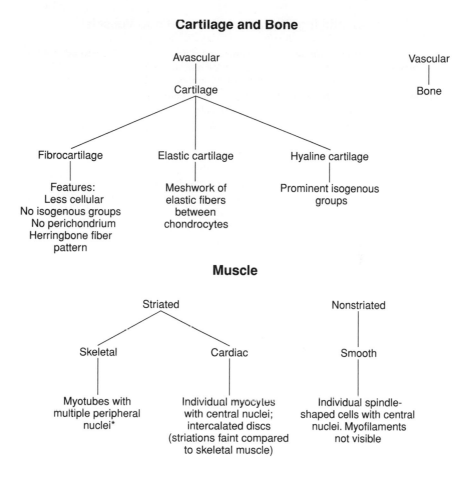

Cartilage and Bone

Avascular

Cartilage

Vascular

Bone

Fibrocartilage

Features:
Less cellular
No isogenous groups
No perichondrium
Herringbone fiber
pattern

Elastic cartilage

Meshwork of
elastic fibers
between
chondrocytes

Hyaline cartilage

Prominent isogenous
groups

Muscle

Striated

Skeletal

Myotubes with
multiple peripheral
nuclei*

Cardiac

Individual myocytes
with central nuclei;
intercalated discs
(striations faint compared
to skeletal muscle)

Nonstriated

Smooth

Individual spindle-
shaped cells with central
nuclei. Myofilaments
not visible

*Students frequently confuse tendon with skeletal muscle because of the banding pattern in highly organized collagen bundles. Look for the nuclei of fibroblasts between the collagen bundles in tendon.

Figure 28-1   Bone/cartilage and muscle strategy.

Identifying Characteristics of Blood Vessels*

	Tunica intima	Tunica media	Tunica adventitia
Large artery		Elaborate elastic fibers throughout	
Medium artery	Prominent internal elastic lamina	Prominent smooth muscle cells	
Arteriole	Visible internal elastic lamina (compare to venule)	Prominent smooth muscle (2–5 layers)	No external elastic lamina
Capillary	No internal elastic lamina	No smooth muscle, pericytes may be present	
Large vein	Difficult to distinguish internal elastic lamina	Thin smooth muscle layer compared to similar size artery	Longitudinal smooth muscle bundles
Medium vein	No visible internal elastic lamina. Valves may be present	Thin smooth muscle layer compared to similar size artery	
Venule	No internal elastic lamina	Thin smooth muscle layer compared to arteriole	
Thoracic duct**	Total disorganization of the vascular wall		

*Most students try to separate arteries and veins on the basis that veins are usually collapsed. The above features are more reliable and easy to learn.

**Lymphatic capillaries have a larger lumen than blood capillaries. They have extremely thin walls and are usually collapsed and irregular in outline. Valves may be present.

Figure 28-2   Blood vessel strategy.

Endocrine Glands

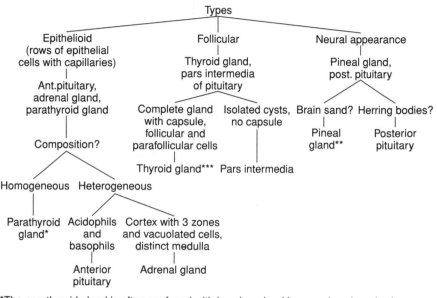

*The parathyroid gland is often confused with lymph nodes. However, lymph nodes have a distinct cortex and medulla while the parathyroid gland is very homogeneous.

**The parathyroid and pineal glands both have a very homogeneous appearance. However, the parathyroid has a definite CT capsule while the pineal gland may contain brain sand.

***The lactating mammary gland is sometimes confused with the thyroid gland. Remember that the mammary gland is an exocrine gland with a duct system.

Figure 28-3   Endocrine strategy.

Exocrine Glands

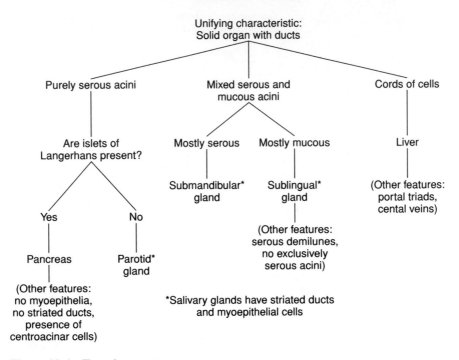

Figure 28-4   Exocrine strategy.

Female Reproductive System

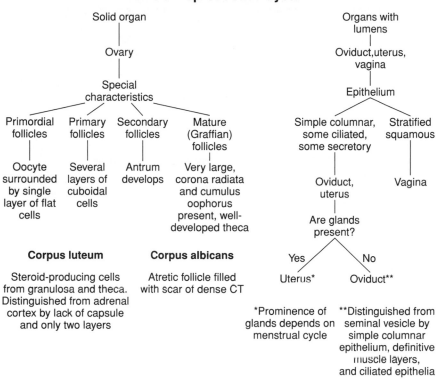

Corpus luteum

Steroid-producing cells
from granulosa and theca.
Distinguished from adrenal
cortex by lack of capsule
and only two layers

Corpus albicans

Atretic follicle filled
with scar of dense CT

Figure 28-5   Female reproductive system strategy.

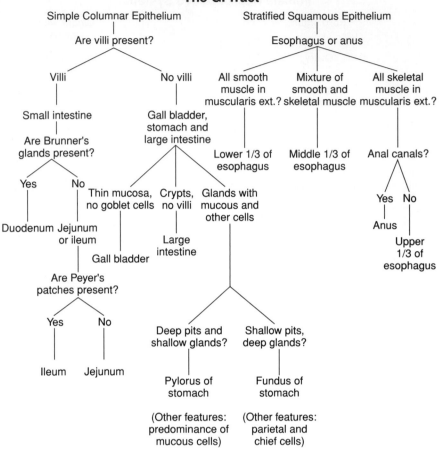

Figure 28-6   Gastrointestinal tract strategy.

Lymphoid Organs

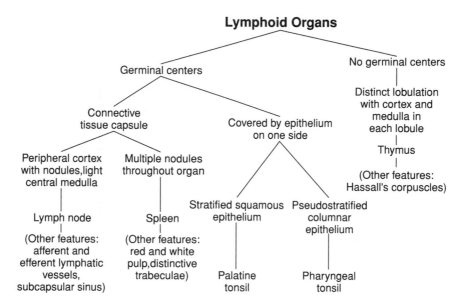

Figure 28-7   Lymphoid strategy.

Male Reproductive System
Epithelium

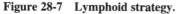

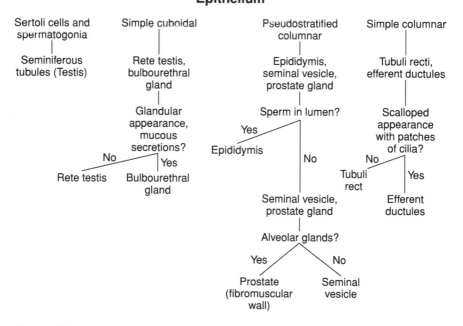

Figure 28-8   Male reproductive system strategy.

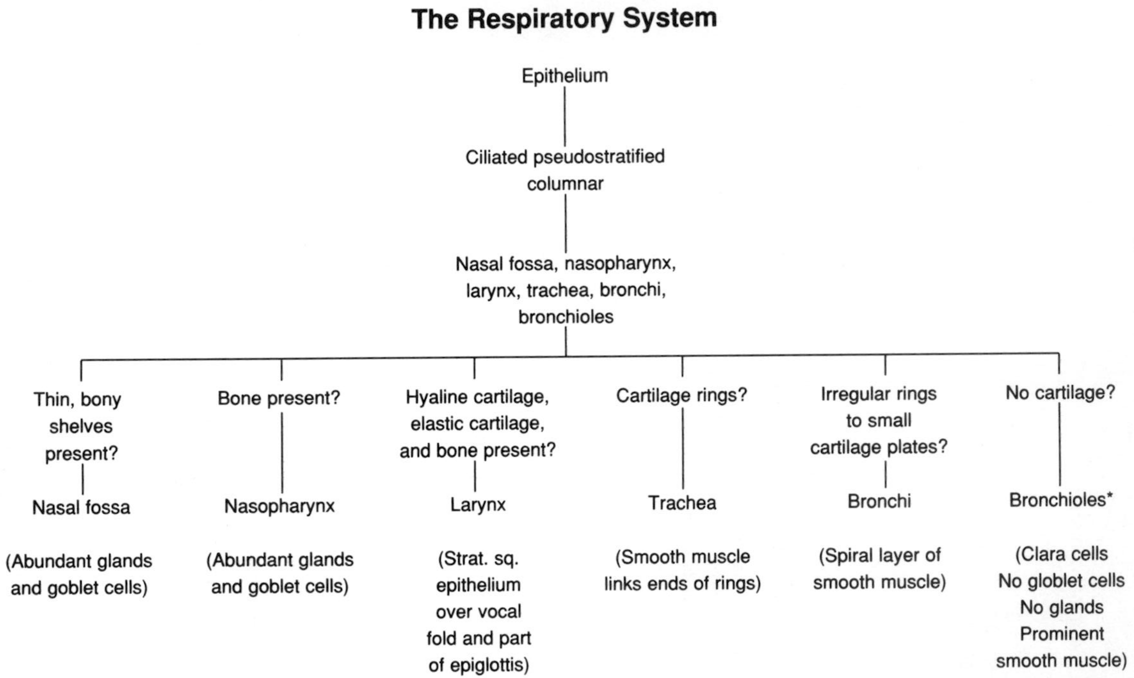

The Respiratory System

Epithelium

Ciliated pseudostratified columnar

Nasal fossa, nasopharynx, larynx, trachea, bronchi, bronchioles

Bone present?

- Thin, bony shelves present?

  Nasal fossa

  (Abundant glands and goblet cells)

- Nasopharynx

  (Abundant glands and goblet cells)

Cartilage rings?

- Hyaline cartilage, elastic cartilage, and bone present?

  Larynx

  (Strat. sq. epithelium over vocal fold and part of epiglottis)

- Trachea

  (Smooth muscle links ends of rings)

No cartilage?

- Irregular rings to small cartilage plates?

  Bronchi

  (Spiral layer of smooth muscle)

- Bronchioles*

  (Clara cells
  No globlet cells
  No glands
  Prominent smooth muscle)

*Terminal bronchioles possess a ciliated, simple columnar epithelium while respiratory bronchioles possess a ciliated, cuboidal epithelium with alveolar pockets.

Alveoli consist of a simple epithelium of Type I and Type II pneumocytes.

Figure 28-9   The respiratory system strategy.

Urinary System

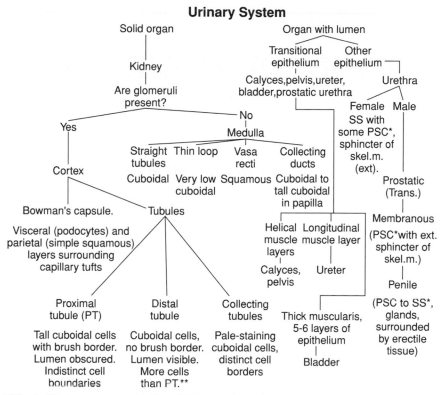

*SS, stratified squamous epithelium; PSC, pseudostratified columnar epithelium

**PT, proximal tubule

Figure 28-10   The urinary system strategy.

TOOLS OF THE TRADE

•

- **Light Microscope**
- **Electron Microscope**
- **Visualization Methods: Stains for Light and Electron Microscopy**
- **Immunohistochemistry**
- **Autoradiography**

•   •   •   •   •   •   •   •   •   •   •   •

LIGHT MICROSCOPE

The light microscope uses a beam of transmitted light to illuminate a tissue section (transillumination). The lenses of the microscope are used to project and enlarge the image.

Essentially, a light beam passes through the tissue. Subsequently, two sets of lenses (objectives and oculars) project and enlarge the image. The objectives project to the oculars, and the oculars project to the eye of the observer. The numbers on the side of the objectives and the oculars followed by an $\times$ represent the magnification of each lens. The total magnifying power of the microscope is calculated by multiplying the magnification of the objective being used by the ocular [e.g., $10\times$ (ocular) multiplied by $4\times$ (objective) results in a total magnifying power of $40\times$]. This example is the typical low magnification of most commonly used microscopes. High magnification is usually about $400\times$ [$10\times$ (ocular) multiplied by $40\times$ (objective)].

Magnification is important, but the key factor for microscopy is the resolution, or resolving power—the ability to distinguish two closely positioned objects as separate.

In the light microscope, 0.2 μm (200 nm) is about the smallest distance at which two individual objects can be distinguished as distinct structures. Although it is not essential to memorize these numbers, it is important to keep in mind that some structures cannot be seen at the conventional light microscopic level. Special light microscopic methods can facilitate visualization. Other forms of light microscopy modify the type of light and the lenses to produce images: fluorescence, polarizing, phase contrast, differential interference, and confocal microscopy all use modified lenses to alter the type of light and images that are created in the light microscope. The limitation still remains about 1000× because of the limited 0.2 mm resolving power. If additional resolving power is desired, the method used is electron microscopy, which provides a different view of tissue.

ELECTRON MICROSCOPE

The electron microscope uses an electron beam in place of a beam of photons (like the light microscope) and electromagnetic lenses instead of glass lenses. There are two kinds of electron microscopy: transmission electron microscopy and scanning electron microscopy.

The most common form of electron microscopy is transmission electron microscopy (TEM), in which an electron beam is passed through an ultrathin tissue section. Some of the electrons in the beam pass through the specimen, whereas others are blocked by the density of the tissue (enhanced by heavy metal staining, discussed in the next section on staining). Variation in density leads to electron-dense and electron-lucent areas. For example, the trilaminar structure of the membrane consists of two electron-dense hydrophilic ends of the phospholipid molecules, whereas the hydrophobic middle portions containing glycerol and fatty acids are electron-lucent. The level of observation visualized in the TEM is known as ultrastructure, or fine structure. The fine structure of a cell can be viewed by an observer on an electron microscope, but most often is recorded as a photographic print.

Electron microscopic tissue preparation is similar to light microscopic tissue preparation, but the tissue samples are smaller and the fixatives and embedding materials are different from light microscopic preparation. The sections are much thinner to optimize resolution and passage of the electron beam through the tissue.

Scanning electron microscopy (SEM) involves visualization of the three-dimensional surface of structures rather than tissue sections. In this case the specimens are fixed and coated with a metal such as gold or platinum, prior to viewing. A secondary electron beam is bounced off the coated surface giving a view of the topography of the surface. Freeze fracture is an important SEM method. Freeze-fracture electron microscopy takes advantage of the ability of biologic materials to fracture along planes of least resistance when force is applied. In the case of cells, this takes place between the leaves of the lipid bilayer. The lipid layer adjacent to the cytoplasm (P face) usually contains numerous particles representing transmembrane proteins, whereas the lipid layer adjacent to the extracellular space (E face) contains only very few particles.

VISUALIZATION METHODS
Stains for light microscopy

Biologic stains take advantage of the chemical nature of cellular constituents such as the basic nature of most proteins and the acidic properties of ribonucleic acids. Some techniques stain many different molecules within a cell and are used for general microscopy, whereas others are more specific.

Hematoxylin and eosin (H&E) is the most commonly used stain for light microscopy.

Hematoxylin is a chemical base which stains acids (base-loving, or basophilic, molecules). Consequently, it binds well to DNA and RNA and results in blue staining of the nucleus. Its companion, eosin, is an acid which binds to bases (acid-loving, or acidophilic, molecules). Since proteins are generally basic, eosin imparts a red-pink stain to the protein-rich cytoplasm.

The periodic acid–Schiff (PAS) technique stains substances which contain glucose such as glycoproteins, mucous carbohydrates, and glycogen.

Periodic acid oxidizes specific groups in glucose residues to aldehydes, which then react with the Schiff reagent to produce a red-violet stain. It is used to detect glycogen deposits and carbohydrate-rich substances such as the mucous contents of goblet cells or the neutral glycoproteins in basement membranes. Alcian blue is another stain which binds to glycosaminoglycans (GAGs) and acidic glycoproteins.

Enzymatic reactions are useful techniques for localizing enzymes such as acid phosphatase in tissues. The resident enzyme hydrolyzes a substrate reagent which reacts with other constituents of the stain to produce an insoluble colored precipitate at the site of the reaction.

Acid phosphatase in tissues hydrolyzes the substrate glycerophosphate to release free phosphate, which reacts with lead nitrate to produce lead phosphate. Lead phosphate, in turn, reacts with ammonium sulfide to produce lead sulfide, an insoluble black precipitate. Enzymatic reactions are especially important because they form the basis for visualization of many immunohistochemical stains (see below).

Toluidine blue is a basic stain commonly used for staining tissues embedded in epoxy resins and for identification of mast cells in paraffin sections.

Toluidine blue provides a useful analogue to low-magnification electron micrographs because the blue staining is directly related to electron density. Toluidine blue is described as a metachromatic stain because it turns tissue a different color than the color of the dye. It is used to identify mast cells, the granules of which turn metachromatic (purple) as they modify and concentrate the dye.

Trichrome stains, as the name implies, contain three dyes which stain different tissue components.

There are several different trichrome stains that are used effectively to stain connective tissue. For the beginning student, use of a trichrome stain in conjunction with a hematoxylin and eosin stain of a similar section facilitates the identification of connective tissue versus smooth muscle.

Stains for electron microscopy

Osmium tetroxide has high affinity for the polar head groups of lipids and serves as both a fixative and a stain for lipid membranes at the EM level.

Although its mechanism of action is yet to be delineated, osmium stabilizes lipid membranes and contributes density which repels electrons.

Heavy metals such as lead (lead acetate) and uranium (uranium acetate) bind to proteins and other cellular constituents. These substances provide a barrier to the electron beam and result in a negative image.

IMMUNOHISTOCHEMISTRY

Antibodies raised against a specific antigen (usually a protein or polypeptide) are obtained from whole animals (polyclonal antibodies), or isolated selected lymphocytes (monoclonal antibodies) and are used to locate the specific antigen in tissue sections (Fig. 29-1).

Antibodies bind to antigens in tissue sections but require signaling systems to make the localization visible.

Various methods have been employed to visualize the antigen-antibody complex in tissue sections at both the light and electron microscopic levels. Early studies utilized enzymes such as peroxidase linked directly to the antibody (direct method). The peroxidase hydrolyzed peroxide in a step which borrowed an electron from a chromogen, leaving a visible precipitate at the site of the reaction (Fig. 29-1). Fluorescent labels such as fluorescein isothiocyanate (FITC) are often substituted for enzymes.

Refinements in the technique eventually led to the indirect method in which a second antibody, raised against the first antibody in a different species (e.g., goat anti-rabbit IgG), carries the detection system and is bound to the first antigen-antibody complex (Fig. 29-1). Currently, the most commonly used enzyme technique uses a three-step approach. After binding the primary antibody to the antigen, the sections are exposed to a secondary antibody labeled with biotin. Biotin has no signal activity itself but strongly binds an avidin-biotin-peroxidase complex (ABC) added in the third step. The peroxidase hydrolyzes peroxide in the presence of a chromogen, as described earlier. This technique provides excellent specificity and signal amplification.

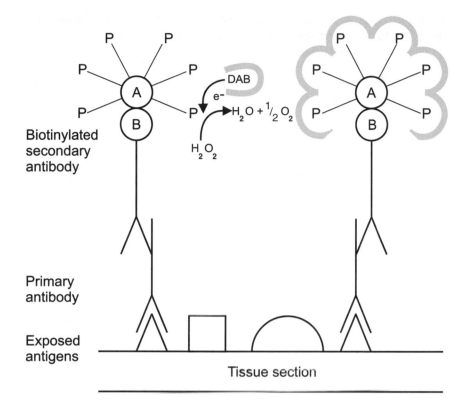

Figure 29-1 Immunohistochemistry with the ABC technique.
The tissue section is reacted with the primary antibody against the antigen of interest (triangle). Subsequent steps are detailed in the text. Other antigens (square, semicircle) do not react. *A*. avidin. *B*. biotin. *DAB*. diamenobenzaldehyde. *P*. perioxidase.

Peroxidase reactions have also been used at the EM level, but the most versatile technique currently employed uses secondary antibodies labeled with gold particles. Since different sizes of gold particles can be used, it is possible to localize several antigens on one specimen.

AUTORADIOGRAPHY

Autoradiography utilizes a radioactive label incorporated into living tissues to localize specific events such as the processes of protein and DNA synthesis.

Labeled precursors injected into an animal are detected in tissue sections by coating the specimen with a photographic emulsion or exposing the section to a photographic film. In essence, the length of exposure to the film or emulsion is akin to setting the shutter speed of a camera. This technique was used to establish the protein synthetic pathway of rough endoplasmic reticulum (RER)–Golgi apparatus–secretory vesicle exocytosis outlined in Chap. 5. Autoradiography is also used to localize specific receptors for neurotransmitters and hormones in tissue sections and isolated cells.

In situ hybridization

Hybridization techniques use strands of DNA, which consist of a sequence of bases complementary to the sequences of DNA or RNA that are to be localized. Both radioactive labels and enzyme techniques can be utilized to visualize the labeled nucleic acid strands either on a film or on tissue sections.

I N D E X

INDEX

Note: Page numbers in italic type refer to figures.

ISBN 0-07-036930-5

90000